HOLT, RINEHART AND WINSTON

Building Formulas

BRITANNICA Mathematics in Context

Britannica

Mathematics in Context is a comprehensive curriculum for the middle grades. It was developed in collaboration with the Wisconsin Center for Education Research, School of Education, University of Wisconsin–Madison and the Freudenthal Institute at the University of Utrecht, The Netherlands, with the support of National Science Foundation Grant No. 9054928.

National Science Foundation

Opinions expressed are those of the authors and not necessarily those of the Foundation

Revision Project

Peter Sickler
Project Director

Teri Hedges
Revision Consultant

Nieka Mamczak
Revision Consultant

Erin Turner
Revision Consultant

Cheryl Deese
MiC General Manager

Vicki Mirabile
Project Manager

2003 Printing by Holt, Rinehart and Winston

ISBN 0-03-071622-5

2 3 4 5 170 05 04 03

The *Mathematics in Context* Development Team

Mathematics in Context is a comprehensive curriculum for the middle grades. The National Science Foundation funded the National Center for Research in Mathematical Sciences Education at the University of Wisconsin–Madison to develop and field-test the materials from 1991 through 1996. The Freudenthal Institute at the University of Utrecht in The Netherlands, as a subcontractor, collaborated with the University of Wisconsin–Madison on the development of the curriculum.

The initial version of *Building Formulas* was developed by Monica Wijers, Anton Roodhardt, and Martin van Reeuwijk. It was adapted for use in American schools by Gail Burrill, Beth R. Cole, and Margaret A. Pligge.

National Center for Research in Mathematical Sciences Education Staff

Thomas A. Romberg
Director

Joan Daniels Pedro
Assistant to the Director

Gail Burrill
Coordinator
Field Test Materials

Margaret R. Meyer
Coordinator
Pilot Test Materials

Mary Ann Fix
Editorial Coordinator

Sherian Foster
Editorial Coordinator

James A. Middleton
Pilot Test Coordinator

Project Staff

Jonathan Brendefur
Laura J. Brinker
James Browne
Jack Burrill
Rose Byrd
Peter Christiansen
Barbara Clarke
Doug Clarke
Beth R. Cole
Fae Dremock
Jasmina Milinkovic
Margaret A. Pligge
Mary C. Shafer
Julia A. Shew
Aaron N. Simon
Marvin Smith
Stephanie Z. Smith
Mary S. Spence

Freudenthal Institute Staff

Jan de Lange
Director

Els Feijs
Coordinator

Martin van Reeuwijk
Coordinator

Project Staff

Mieke Abels
Nina Boswinkel
Frans van Galen
Koeno Gravemeijer
Marja van den Heuvel-Panhuizen
Jan Auke de Jong
Vincent Jonker
Ronald Keijzer
Martin Kindt
Jansie Niehaus
Nanda Querelle
Anton Roodhardt
Leen Streefland
Adri Treffers
Monica Wijers
Astrid de Wild

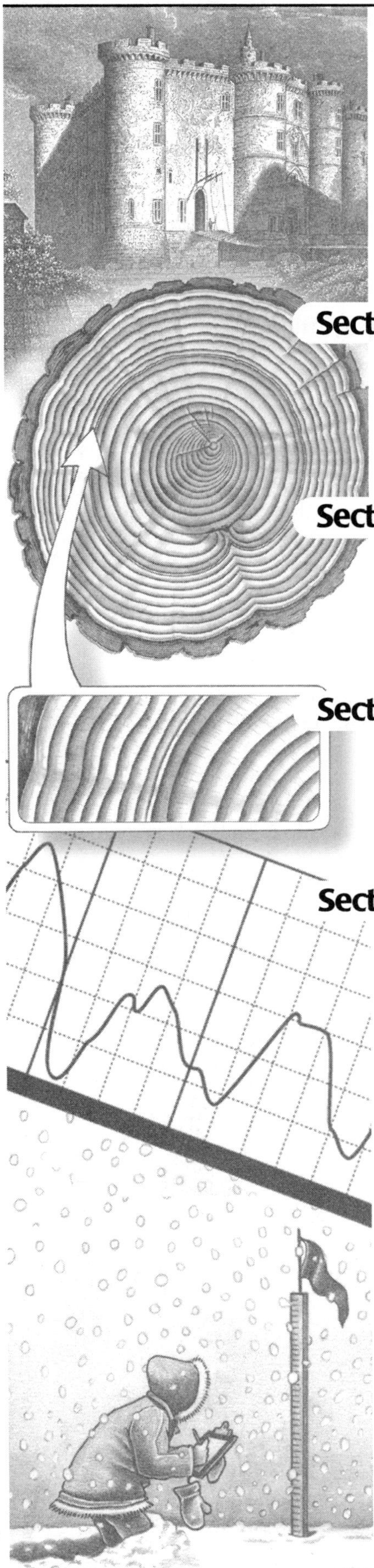

Table of Contents

Dear Student,

Welcome to *Building Formulas.*

Throughout this unit, you will study many kinds of formulas.

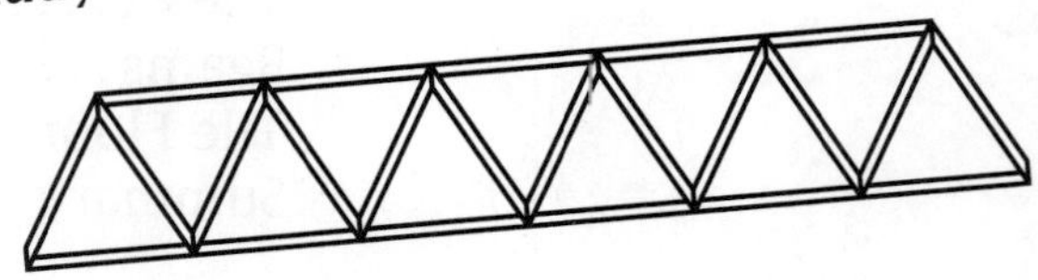

You will first learn to identify the parts that make up a formula and then create your own formulas. You will help with the construction of a movie set by creating a formula for determining the numbers of gray and white tiles and metal rods needed.

Do you know how to find the radius of a circle if the area is 154,000 square meters? In order to visualize the area covered by one *Armillaria bulbosa* fungus, you will learn how to work with square roots and discover a formula for finding the radius of a circle.

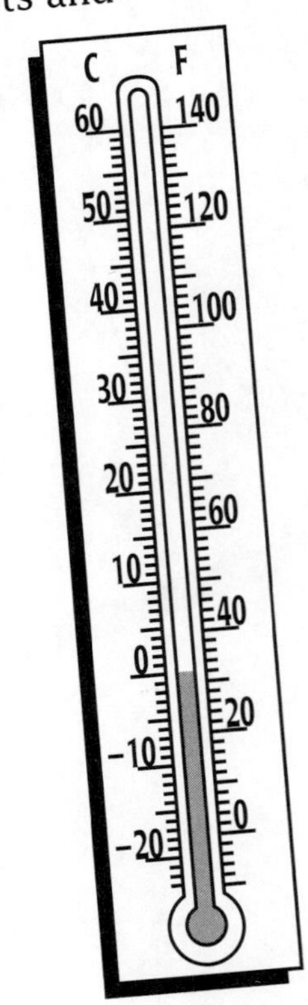

You will also work with formulas that someone else created, such as a formula for converting from degrees Celsius to degrees Fahrenheit and formulas that archaeologists use to recreate ancient Egyptian drawings.

Upon completing this unit, you should understand the meanings of the different parts of a formula, how to graph formulas, how to rewrite formulas, and how to use formulas that you find.

Sincerely,

The Mathematics in Context Development Team

A. PATTERNS

Tiles

Urvashi is designing a set for a movie scene that takes place outside a mansion. The mansion is surrounded by a large garden decorated with plants, tiled paths, sculptures, and patios.

She wants to design tile patterns for different lengths of garden paths. She decides to use square tiles in two different colors.

Using one pattern, Urvashi draws paths that have different lengths.

To make it easy to refer to a path, Urvashi assigns each path a number.

1. a. What do the numbers represent?

b. Using different-colored paper or plastic squares, lay out some other lengths of paths that have this pattern. For each example, write down the path number. Describe the patterns in the numbers of tiles for different paths.

Path Number 3

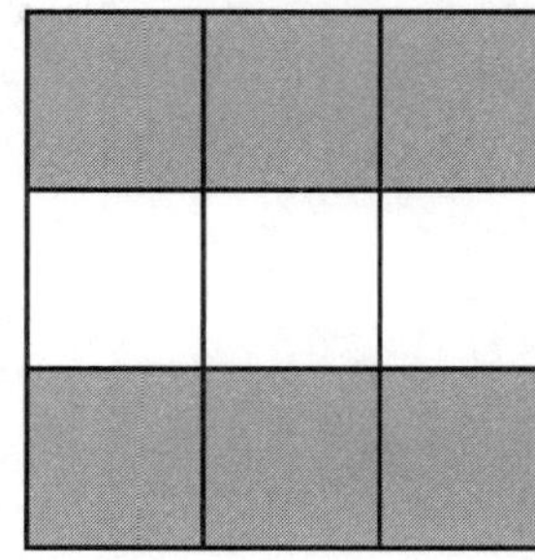

Path Number 4

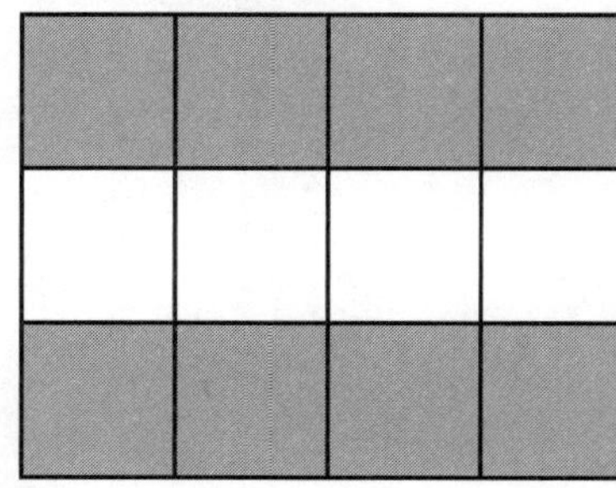

Path Number 5

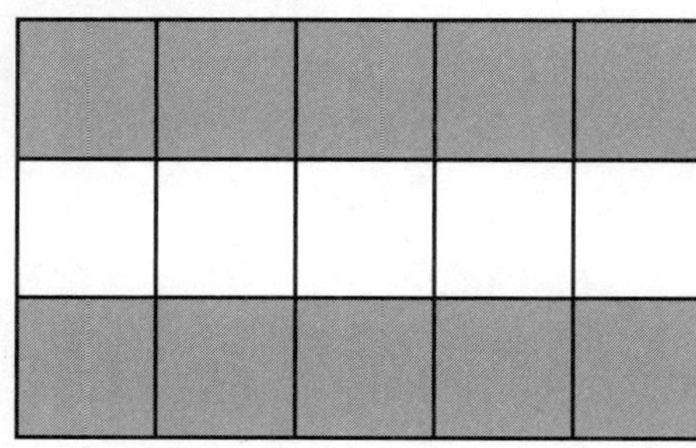

Urvashi wants to make the pattern more interesting and decides to add a column of tiles at the beginning and at the end.

Path Number 3

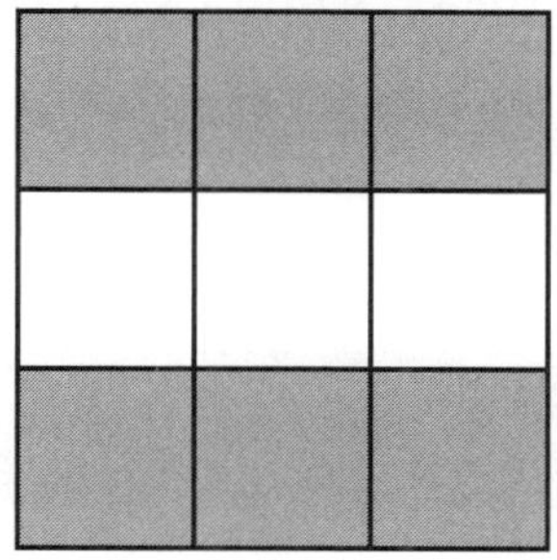

The picture on the right shows how path number 3 is changed.

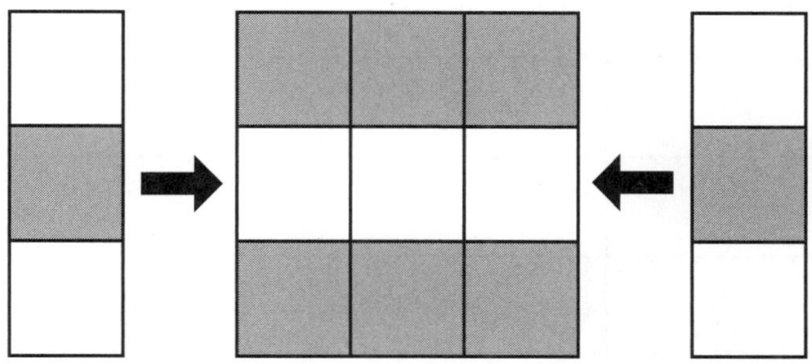

New path number 3 is now five tiles long.

New Path Number 3

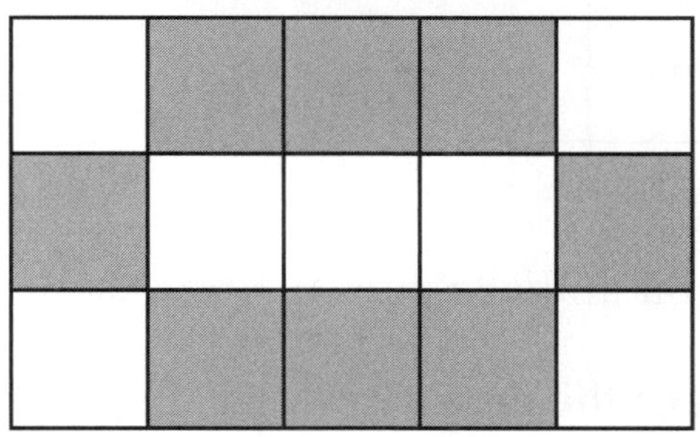

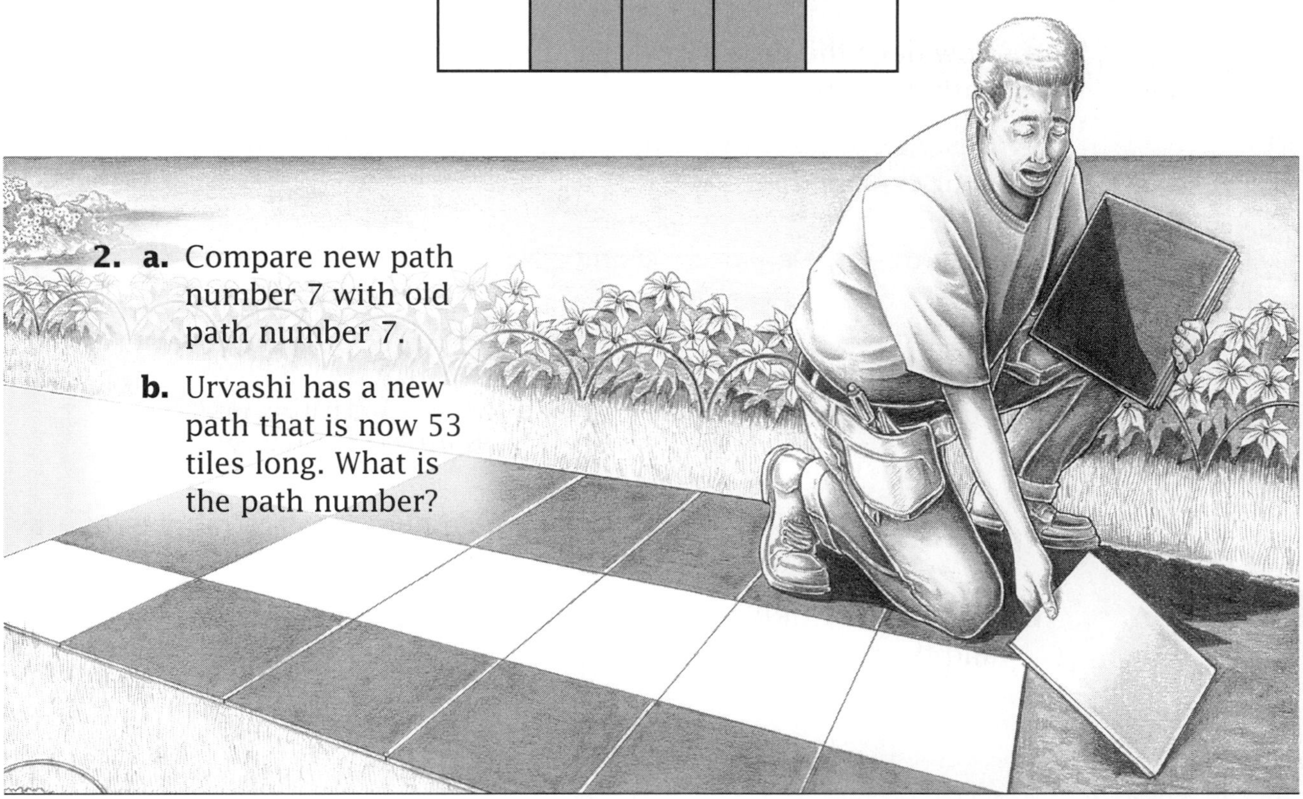

2. a. Compare new path number 7 with old path number 7.

b. Urvashi has a new path that is now 53 tiles long. What is the path number?

3. Use your tiles to make some examples of Urvashi's new design.

Urvashi makes a table so she can easily find the number of tiles needed for each new path design.

Path Number	Number of Gray Tiles	Number of White Tiles	Total Number of Tiles
1			
2			
3	8	7	15
4			
5			
6			
7			

4. Copy and fill in the table.

5. **a.** How does the number of gray tiles change when you go from one path number to the next?

b. Describe at least two other patterns in the table. Use drawings in your explanation.

You can also describe a pattern using a NEXT-CURRENT formula.

The NEXT-CURRENT formula describing the number of gray tiles is:

Number of gray tiles in the NEXT path =
Number of gray tiles in the CURRENT path + _?_

6. **a.** Finish the NEXT-CURRENT formula that describes the number of gray tiles. It can be shortened to this: NEXT gray = CURRENT gray + _?_

b. Write NEXT-CURRENT formulas to describe the number of white tiles and the total number of tiles.

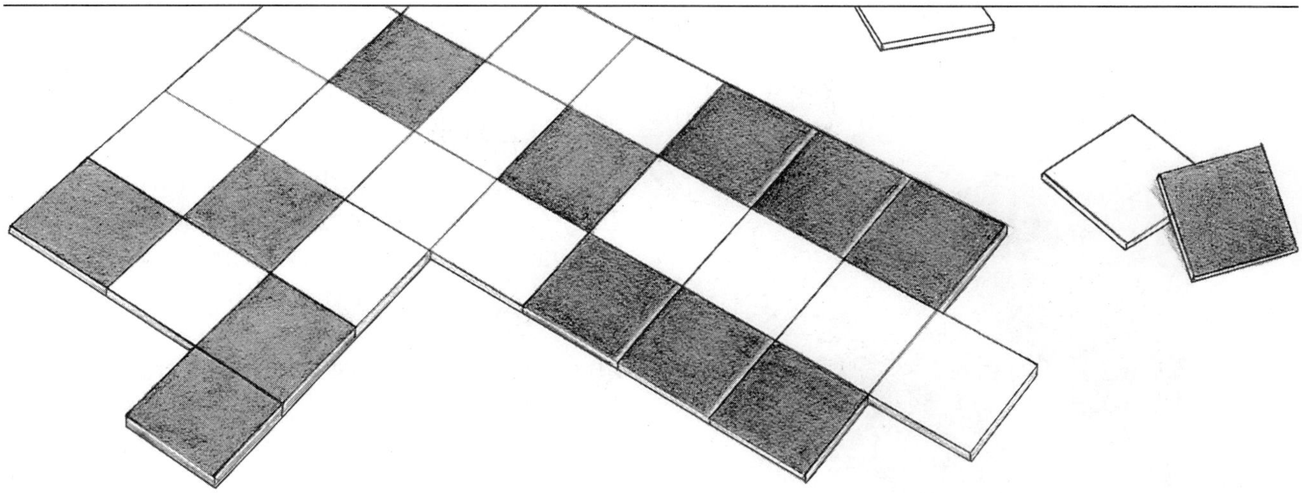

Urvashi makes a model for path number 10 and finds that she needs 14 white tiles.

7. a. Use a NEXT-CURRENT formula to find out how many white tiles she would need for path number 15. Explain how you did this.

b. Find the numbers of white and gray tiles Urvashi would need for path number 30. Explain how you found your answers.

To find the number of gray tiles for path number 30, Urvashi did not want to use the NEXT-CURRENT formula because she did not want to calculate the numbers of tiles, step-by-step, for all the paths before it.

Stu, one of her coworkers, says that figuring the number of gray tiles for path number 30 is easy. All you have to do is calculate $4 + 29 \times 2$.

8. a. Check that Stu's calculation gives the correct number of gray tiles for path number 30.

b. Explain how you think Stu came up with his method.

c. Would the same process work for path numbers other than 30? If so, how?

By looking at examples of different path numbers, Urvashi discovers a relationship that allows her to find the number of white tiles if she knows the path number. She uses P to stand for the path number and W to stand for the number of white tiles. She then writes a *direct formula* in arrow language:

$$P \xrightarrow{+4} W$$

9. **a.** Use a drawing to help explain the above rule.

b. Rewrite the arrow string as a direct formula with an equal sign.

10. **a.** Write a direct formula to find the number of gray tiles for each path number.

b. Explain your formula with a drawing.

c. Compare your rule to Stu's rule.

11. Use the direct formulas to check your answers to problems **7a** and **b** on page 5.

12. Compare the use of NEXT-CURRENT formulas to the use of direct formulas. Explain the advantages and disadvantages of each.

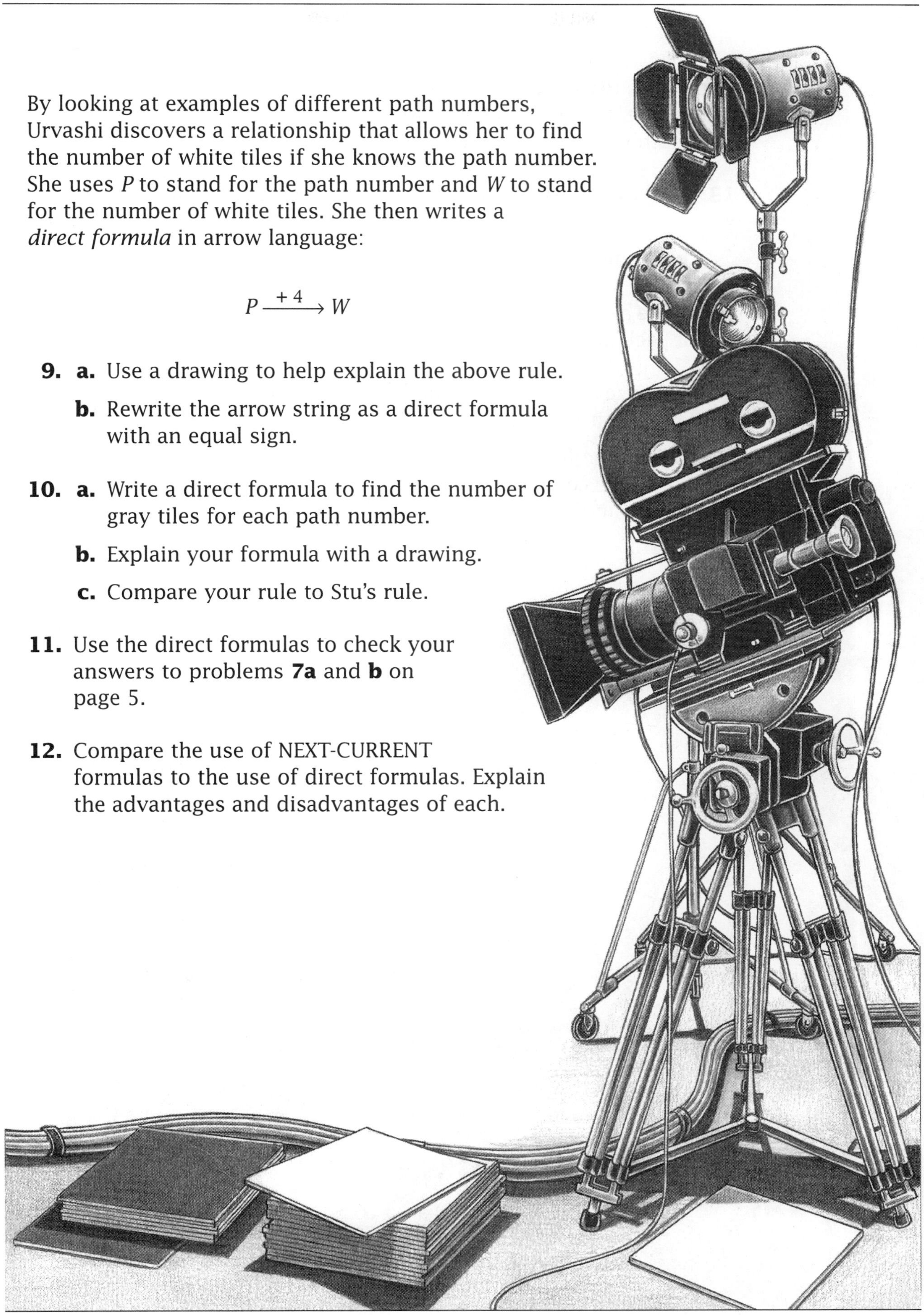

Jim Shew, the budget director for the film, needs to find the cost of the tiles Urvashi will use. Because white and gray tiles are the same price, he needs to know only the total number of tiles for each path number.

13. **a.** Write at least two different ways to find the total number of tiles needed for path number 13.

b. Would these methods work for different path numbers?

Jim wrote a direct formula to calculate the total number of tiles *(T)* for each path number *(P)*.

14. **a.** Write a formula you think Jim might have come up with. Explain the formula.

b. Use the formula to calculate the total number of tiles the movie company would need for path numbers 15, 23, 92, and 93. Show your calculations.

Urvashi and Jim wonder whether any other formulas could be used to find the total number of tiles (T) for each path number (P).

Urvashi suggests the following formula:

$$T = (P + 2) + (P + 2) + (P + 2)$$

15. Will this formula also give the total number of tiles for each path number? Use a drawing to explain your answer.

Jim says this is the same formula as $T = 3 \times (P + 2)$, which can also be written as $T = 3(P + 2)$.

16. a. What did Jim do to get this formula?

b. Find another way to write this formula.

After meeting with the film's director, Urvashi tells Jim they are going to use the new design for the paths. They order 101 tiles.

Jim wonders, "Does a path exist that would require exactly 101 tiles?"

17. a. What is the answer? Explain.

b. What path number contains exactly 54 tiles?

BEAMS

Construction work has begun on a large building that will be used for part of the movie set. The framework consists of metal beams on concrete columns.

Beams can have different lengths. The length of a beam is the number of rods along the underside.

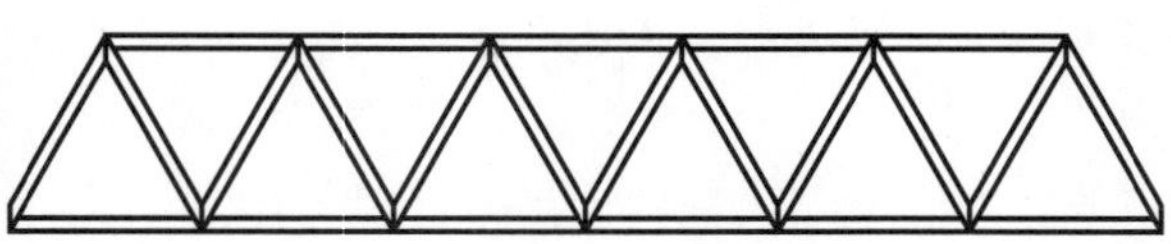

18. Why is the beam on the left considered to be of length 6?

To make the building for the movie set, three beams are put together as pictured below.

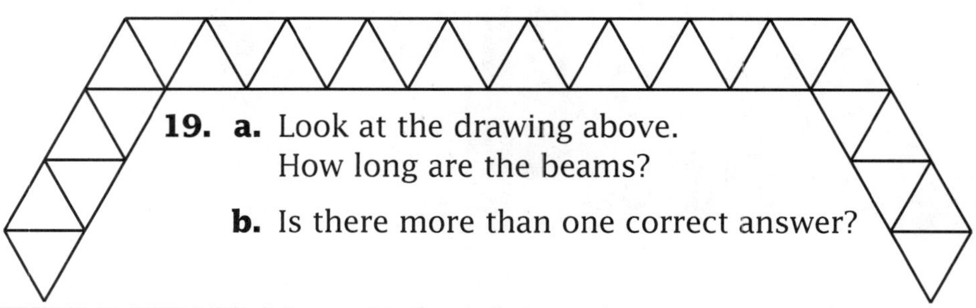

19. a. Look at the drawing above. How long are the beams?

b. Is there more than one correct answer?

In the factory where these beams are made, people want to be able to quickly calculate how many rods are needed for different-length beams.

One way to find the number of rods needed is to look at beams of different lengths.

	Length of Beam	Number of Rods
	1	3
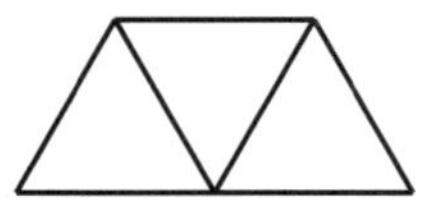	2	
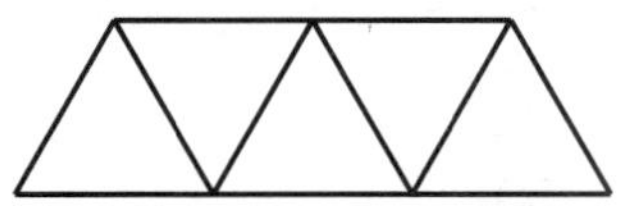	3	
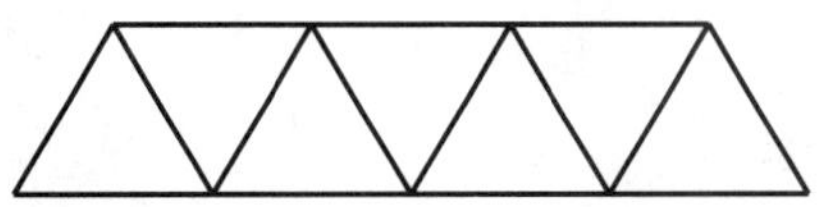	4	

20. a. Copy and complete the above table in your notebook.

b. Explain how you found the numbers to fill in the table.

c. Describe some patterns you can see in the table.

d. Add some entries to the table. Make drawings to check your table entries.

Each time the length of the beam increases by one, the same number of rods is added.

21. a. How many rods are added each time the beam length increases by one?

b. Write a NEXT-CURRENT formula that describes how the number of rods changes from one row to the next.

c. When might you use this NEXT-CURRENT formula?

Jim is interested in finding the total number of rods needed for different lengths of beams.

22. a. If he wants to find out how many rods are needed for a beam of length 12 and one of length 56, could he use a NEXT-CURRENT formula? Explain.

b. Use toothpicks to build a beam of length 6.

c. Find a direct formula that gives the number of rods (R) needed to build a beam of any length (L).

The people who work for Jim want to help him figure out how many rods are needed for different-length beams. Each of them comes up with a formula that they think will work.

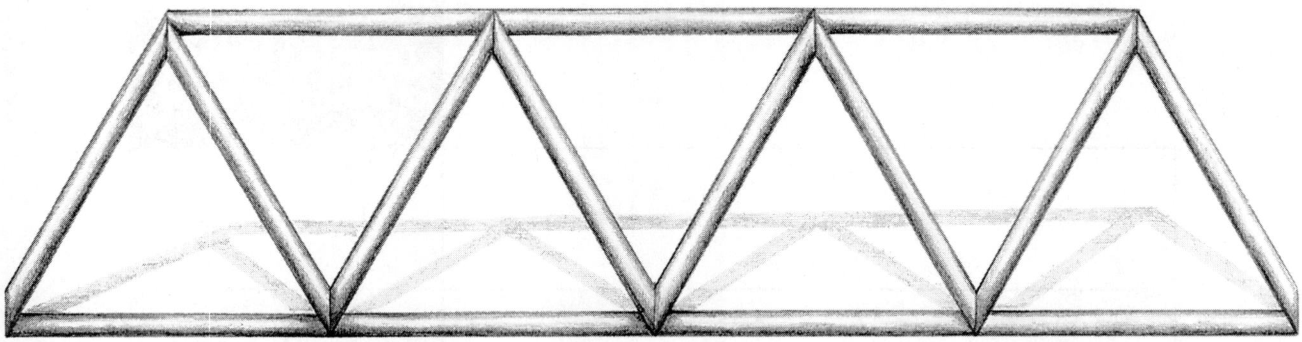

Angelina suggests:	David suggests:	Maria suggests:
$R = L \times 3 + (L - 1)$	$R = L + (L - 1) + 2L$	$R = 3 + (L - 1) \times 4$

Jim is impressed that all of these people want to help, but he wonders whether or not all of these formulas are right.

23. **a.** Check all of the formulas for lengths 7, 15, and 68 to see if they give the same numbers of rods.

b. Are all of the formulas the same? Explain.

c. Compare these formulas with the one you found for problem **22c.**

Jim is not convinced that all three formulas will give the correct results for all lengths. He decides that if he can figure out where the formulas came from, he will have a better idea about whether or not they will work for all lengths.

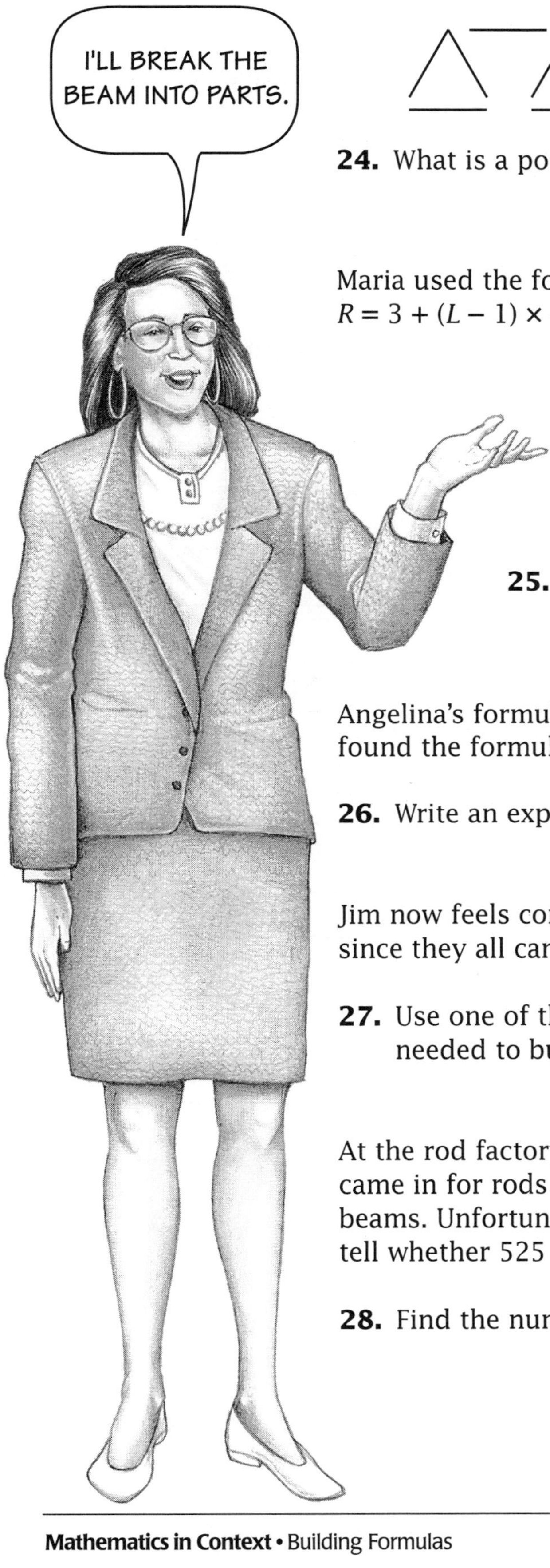

David used the following picture to explain his formula, $R = L + (L - 1) + 2L$:

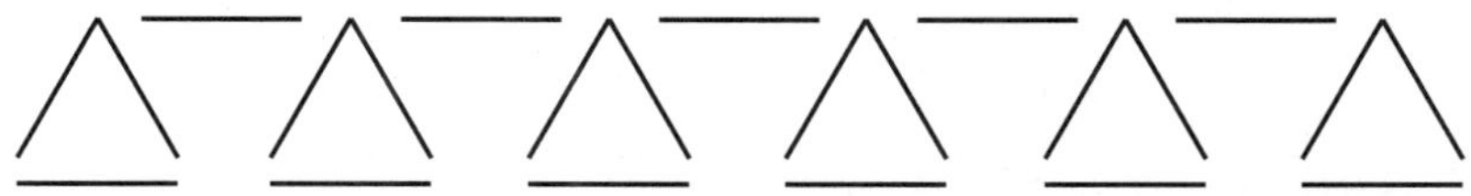

24. What is a possible explanation for David's formula?

Maria used the following picture to describe her formula, $R = 3 + (L - 1) \times 4$.

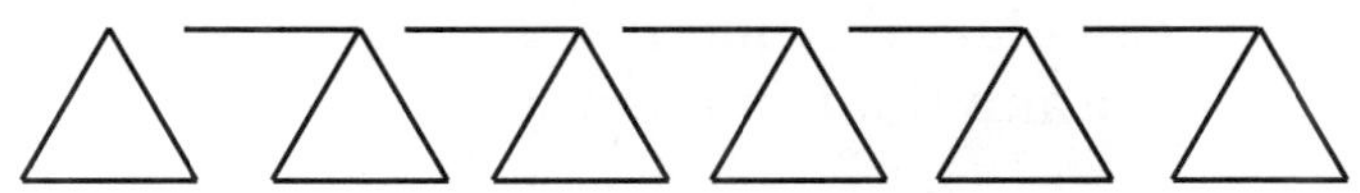

25. Finish Maria's explanation. Explain how the above drawings are related to her formula.

Angelina's formula is $R = L \times 3 + (L - 1)$. She tells Jim that she also found the formula by breaking the beam into parts.

26. Write an explanation for Angelina's formula. Use drawings.

Jim now feels confident using any of the formulas. He figures that since they all came from the same picture, they all must work.

27. Use one of the formulas to calculate the number of rods needed to build the movie set shown on page 10.

At the rod factory, many of the orders come in by fax. An order came in for rods to make a building with seven equal-length beams. Unfortunately, the fax was hard to read, and no one could tell whether 525 or 532 rods had been ordered.

28. Find the number that was ordered.

Tile Floor Design

Terry has to design a tile floor for another scene in the movie.

Her design has a gray square in the middle and a white border around it. These floors can be different sizes. Four different sizes are shown on the right.

Floor Number 1

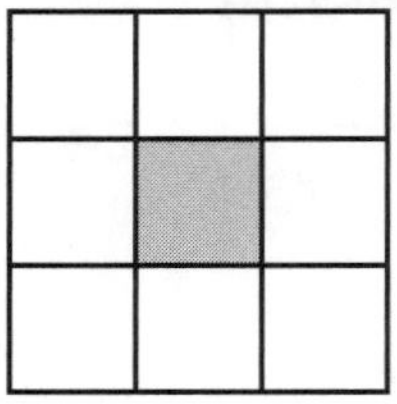

Floor Number 2

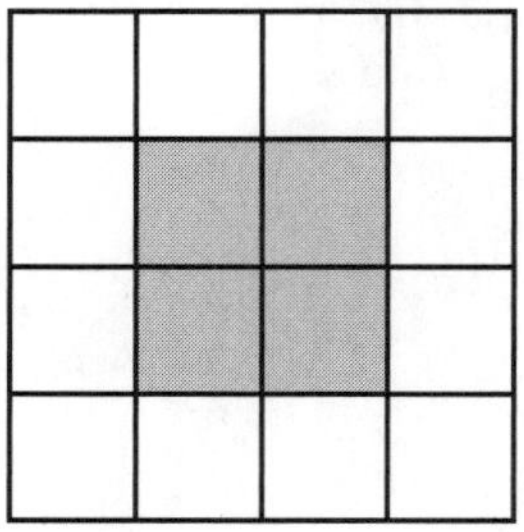

Floor Number 3

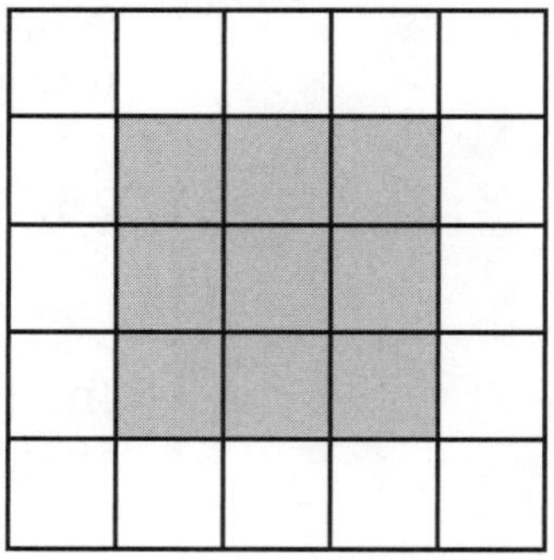

Floor Number 4

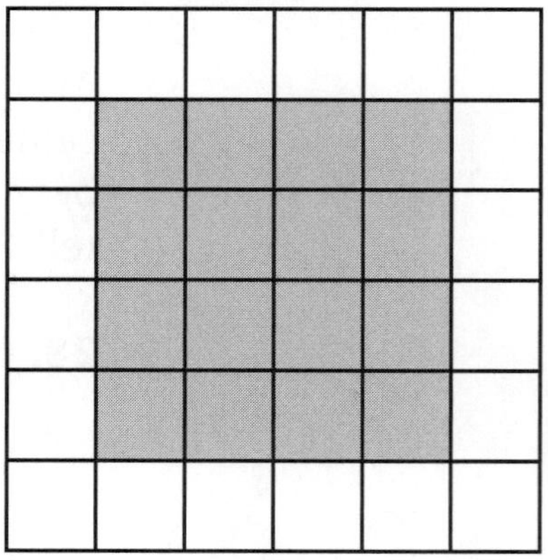

29. a. Study the design. Find a direct formula to calculate the number of gray tiles (G) needed for any floor number (F).

b. Find a direct formula for the number of white (W) tiles in any floor number.

30. Use drawings to explain that either of the following formulas for the total number of tiles will give the correct result:

$$T = F \times F + 4F + 4$$

or

$$T = (F + 2) \times (F + 2)$$

Terry has a total of 144 tiles that she is going to use to build one of these square floors.

31. a. What floor number is she going to build? Explain.

b. How many gray and how many white tiles does she have? Explain.

Summary

When solving problems about designs you can:

- draw some examples of the design,
- make a table and look for patterns,
- express the patterns as formulas.

You have seen two different types of formulas that can be used to describe a pattern:

- a NEXT-CURRENT formula, going step–by–step
- a direct formula, working directly with the pattern number

Different direct formulas can be found to describe the same rule or pattern. You can check to see whether or not different formulas give the same result by connecting each formula to the same pattern using a drawing.

For example, $R = 3L + (L - 1)$ is the same as $R = 4L - 1$ because they both describe the pattern below.

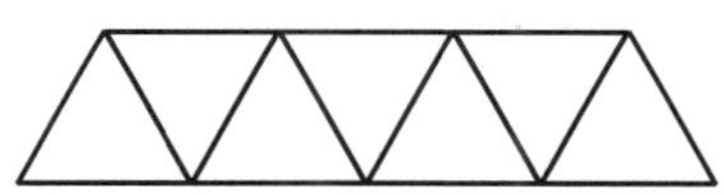

$R = 3L + (L - 1)$ can be connected to the pattern with the following drawing:

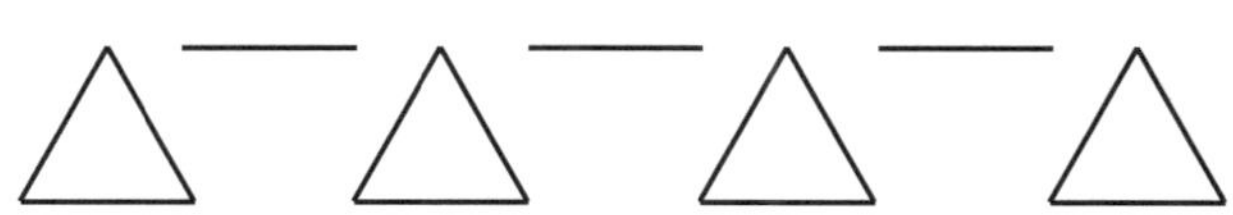

$R = 4L - 1$ represents the same pattern, as shown in the following drawing:

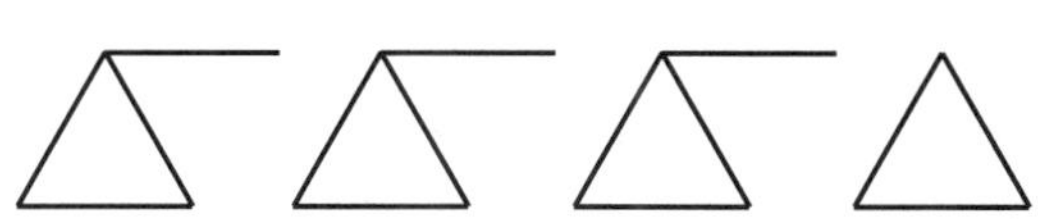

Summary, continued

If the meaning is clear, you can leave out the "×" when multiplying. $T = 4 \times L - 1$ is the same as $T = 4L - 1$.

Also, when you are adding or multiplying, you can change the order. $T = (P + 1) \times 3 + 2$ is the same as $T = 3(P + 1) + 2$.

Summary Questions

32. Explain when you would use a NEXT-CURRENT formula and when you would prefer a direct formula.

33. Here are three formulas:

$$T = P + 2 + P$$

$$T = 2P + 2$$

$$T = 2(P + 1)$$

Show that these formulas describe the same pattern.

BRICKS

Many buildings are made out of bricks. Bricks can be laid in different patterns.

The windows, doors, arches, or even edges of a building may have different brick designs.

You can see different brick patterns on the buildings in these photographs.

Here are some diagrams of brick patterns.

Row 1

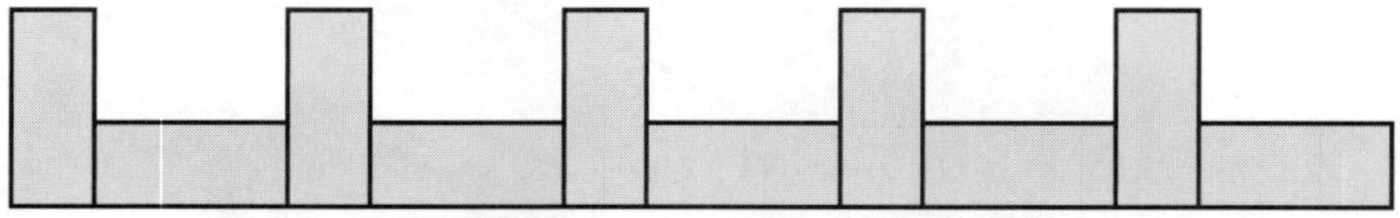

Row 2

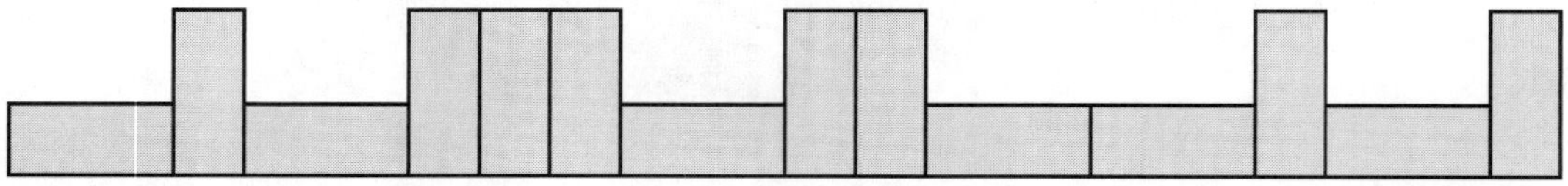

Row 3

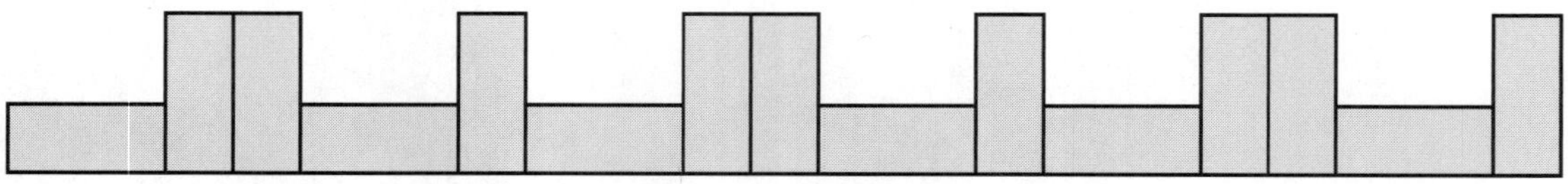

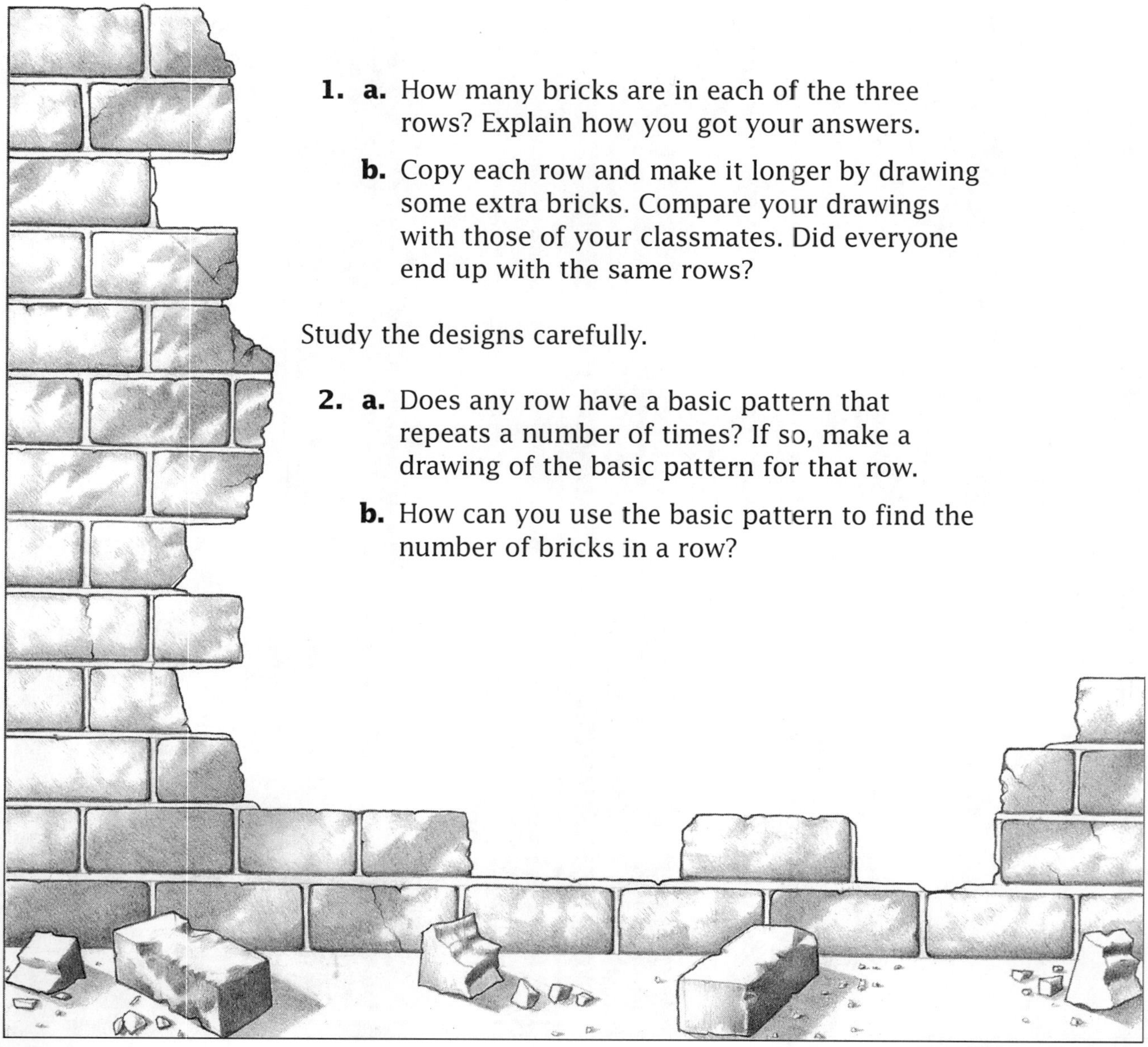

1. **a.** How many bricks are in each of the three rows? Explain how you got your answers.

   b. Copy each row and make it longer by drawing some extra bricks. Compare your drawings with those of your classmates. Did everyone end up with the same rows?

Study the designs carefully.

2. **a.** Does any row have a basic pattern that repeats a number of times? If so, make a drawing of the basic pattern for that row.

   b. How can you use the basic pattern to find the number of bricks in a row?

Bricklayers often use a variety of basic patterns in buildings. One way of creating a pattern is to use standing and lying bricks. A bricklayer could describe such patterns using the letter *S* for bricks standing upright and the letter *L* for bricks lying flat.

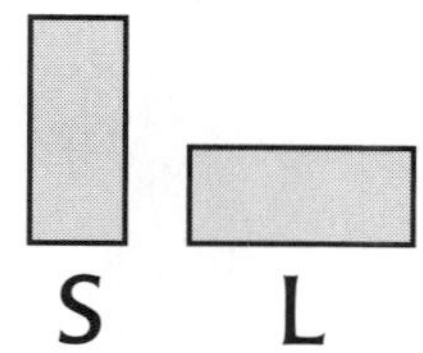

The table below was made by a new bricklayer to help him remember some basic patterns.

3. Use the copy of the table on **Student Activity Sheet 1** to fill in the missing information.

Name	Drawing	String
Classic		*SLSLS*
Exotic	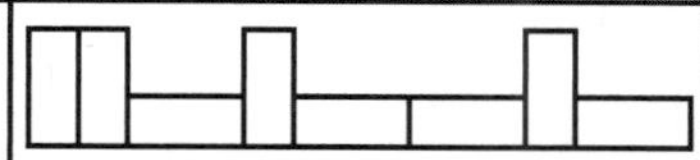	
Modern	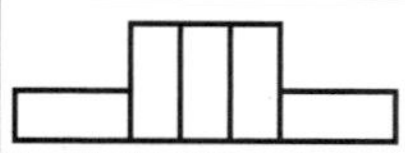	

4. a. On the last line of **Student Activity Sheet 1,** fill in a basic pattern of your own and give it a name.

b. If you were to repeat your basic pattern four times, how many bricks would you need? How did you find this number?

c. Draw or describe what this row would look like.

If a person is building a brick row, it may be important to know not only how many bricks are needed, but also how long (in centimeters or meters) the row will be. Even though you do not know the sizes of the above bricks, you can still make some true statements about the lengths of the basic patterns.

5. Write down two true statements comparing the lengths of the basic patterns shown in the table.

The Classic

Ms. Fix would like to have a brick border for one side of her garden. She has chosen the Classic pattern. In order to make the border long enough, she repeats the pattern four times.

6. a. Describe or draw what Ms. Fix's brick border will look like.

b. How many bricks does she need? Write down your calculations.

Ms. Fix is thinking about using some bricks that she has left over from another project. She has to choose between yellow and gray bricks. They have the following measurements.

7. a. How long would her brick border be if she used only yellow bricks? only gray? Write down your calculations.

b. Did everyone in your class make the calculations in the same way? Explain.

Ms. Fix visits a brickyard to look at some other bricks for her border. She decides to use a formula for finding the length of the basic classic pattern. To use the formula, Ms. Fix needs to know the length of the lying (long) side of the brick (L) and the length of the standing (short) side of the brick (S). This is the formula she uses:

$$\text{Length of Classic} = 3S + 2L$$

8. **a.** Explain the formula.

b. Write formulas for the other basic patterns shown in the table on **Student Activity Sheet 1.**

Ms. Fix is interested in a formula that will give the total length for the brick row. (Remember that she wants to repeat the basic Classic pattern four times.)

She writes:

$$\text{Total Length} = 4 \times \text{Length of Classic}$$

Later she realizes she can also use this formula:

$$\text{Total Length} = 4(3S + 2L)$$

9. Explain how both of the formulas work.

Ms. Fix wants a formula that shows how many short sides of a brick and how many long sides are used in the total length.

10. Write such a formula for Ms. Fix. (*Hint:* This formula should have no parentheses.)

More Brick Rows

The Yun family has designed a brick row for their garden using a basic pattern of their own. Below you see what the row will look like.

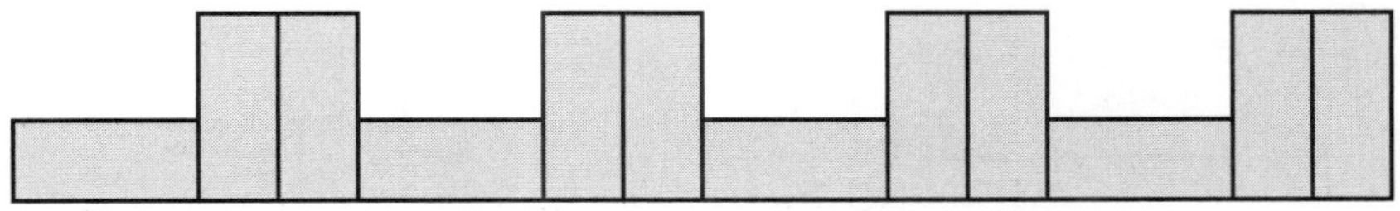

11. a. Describe the basic pattern of this row.

b. Use the letters L and S to write a formula to calculate the length of the basic pattern.

c. Write two formulas you can use to calculate the length of the row shown above. Make one formula with parentheses and one without.

d. Describe how you think the formula with parentheses can be rewritten as a formula without parentheses.

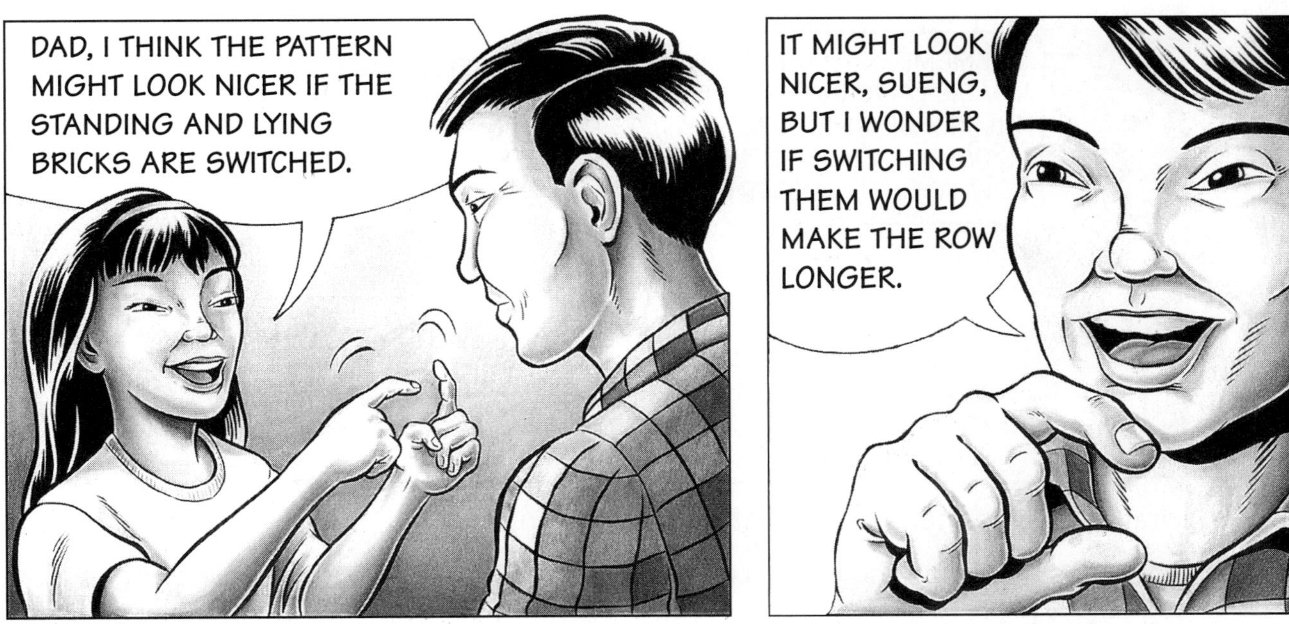

12. a. Does the row's length change if Sueng's idea is carried out? How can you tell?

b. Write formulas for the length of the new basic pattern and for the length of the whole row.

Ms. Peterson is fixing up her house. She goes to a hardware store to buy some supplies.

13. **a.** How can Ms. Peterson use the formula to find the total number of bricks she needs?

b. Draw one possible design for the brick row that Ms. Peterson is repairing.

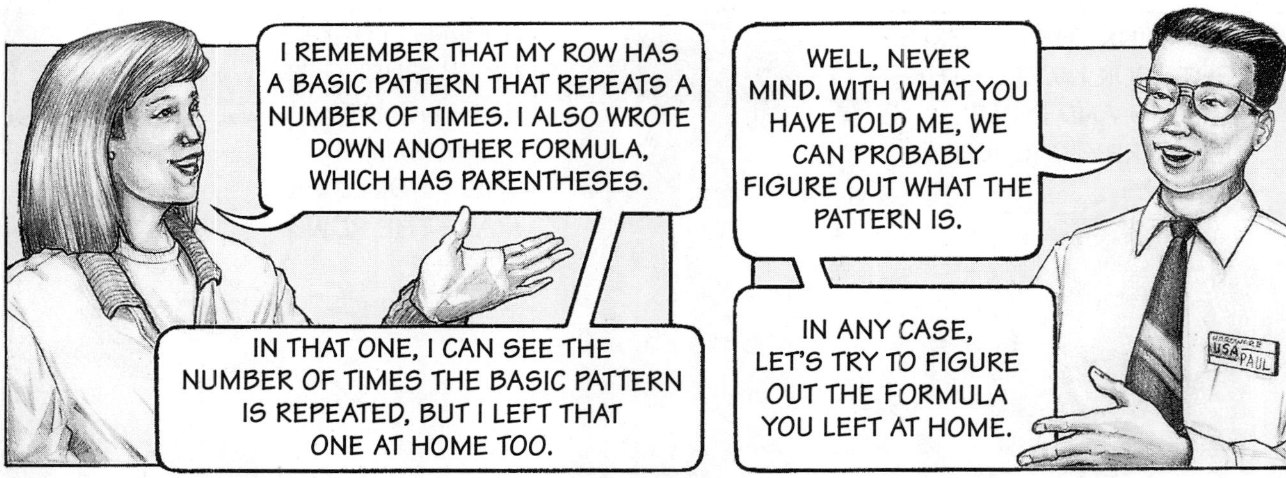

14. **a.** Draw a basic pattern that fits Ms. Peterson's row.

b. Are you sure this is the basic pattern Ms. Peterson has in mind? Explain your answer.

c. Using the basic pattern and parentheses, write a formula to calculate the length of the row Ms. Peterson is restoring.

d. Explain how this formula is related to the formula Length = $15S + 10L$.

15. Could Ms. Peterson's brick row have a basic pattern that occurs four times? Explain why or why not. Support your explanation with drawings.

Another formula for a brick row is:

$$\text{Length} = 3(2S + 4L)$$

16. a. Draw a brick row that fits this formula.

b. Write a formula without parentheses for the length of the row you drew. Explain how you found the numbers for the formula.

You may have noticed that there is a rule you can use to change a formula with parentheses into one without parentheses. The formula Length = $4(2S + 5L)$ can be rewritten as Length = $8S + 20L$.

17. a. Where do the 8 and the 20 come from in the second formula above?

b. If you start with the formula Length = $8S + 20L$ what do you have to do to rewrite it as Length = $4(2S + 5L)$?

Summary

If you have to repeat a calculation over and over, using a formula can be helpful.

In this section, you have seen different formulas for calculating the lengths of brick rows, formulas with and without parentheses.

A formula with parentheses can be rewritten without parentheses. For example, $2(4S + 3L)$ is the same as $8S + 6L$.

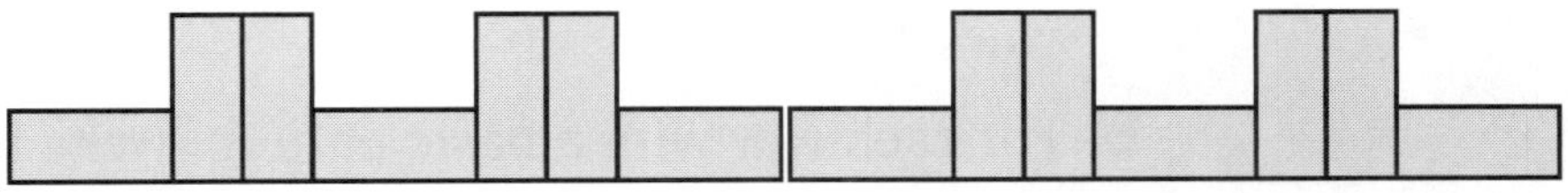

A formula without parentheses can sometimes be rewritten with parentheses. For example, $6S + 3L$ is the same as $3(2S + 1L)$.

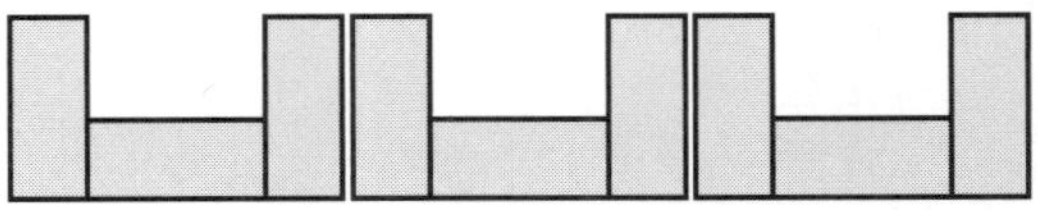

Summary Questions

18. **a.** Describe or draw a brick row that would fit the formula Length = $4(2L + 3S)$.

b. Rewrite this as a formula without parentheses.

The following formulas are used to calculate the lengths of different brick rows. All of the brick rows except one have a repeating basic pattern.

Row 1: Length = $12L + 8S$

Row 2: Length = $10L + 15S$

Row 3: Length = $21L + 27S$

Row 4: Length = $9L + 13S$

Row 5: Length = $18L + 12S$

Row 6: Length = $18L + 9S$

19. **a.** Which of the six rows cannot have a repeating basic pattern? Explain.

b. Row 5 can have two different repeating basic patterns. What are the formulas for the lengths of these basic patterns?

Here are three formulas for the lengths of different basic patterns:

Basic Pattern I: Length = $7L + 9S$

Basic Pattern II: Length = $3L + 2S$

Basic Pattern III: Length = $2L + 3S$

20. **a.** Which basic patterns (I, II, or III) can be used for each of rows 1–6?

b. For each row with a basic pattern, write how many times the basic pattern is repeated.

HUMONGOUS FUNGUS

On April 3, 1992, biologists in northern Michigan reported that they had found what they believed to be the world's largest living organism, an *Armillaria bulbosa* fungus. The fungus is made up of an underground network of hair-like strands from which sprout mushrooms (see picture below). The mushrooms feed on dying trees. Scientists collected *Armillaria bulbosa* mushrooms in a 154,000-square meter area and, through genetic testing, concluded that they were all part of the same fungus.

It is difficult to imagine how large an area 154,000 square meters is.

1. a. Find some combinations of length and width measurements for an area of 154,000 square meters.

b. Name something that covers about the same area as the fungus.

Shown above is the cleared site in a hardwood forest of red pine where the *Armillaria bulbosa* was found.

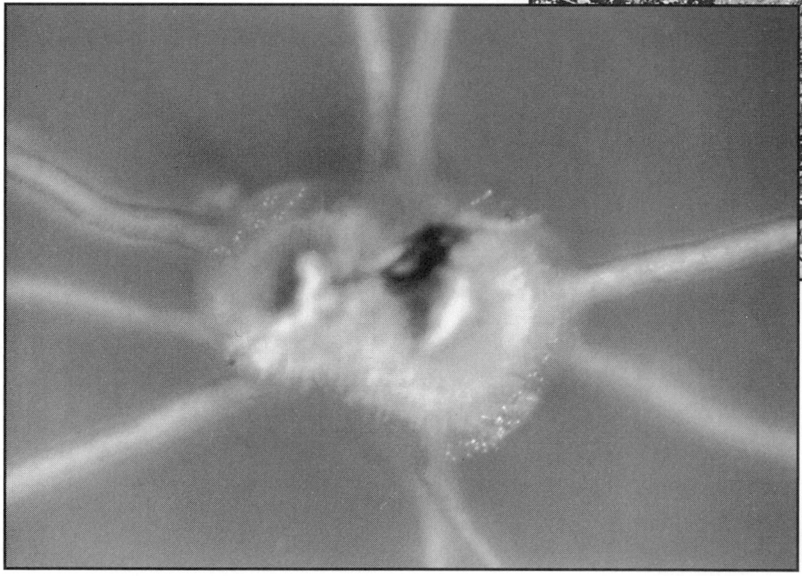

Shown at left is the *Armillaria* fungus rhizomorphs, or "tentacles," magnified ten times.

Square Numbers

To get a better idea of how large the fungus is, you can explore the connection between the length of a side and the area of a square. Below is a 10 × 10 grid that has a "growing white square."

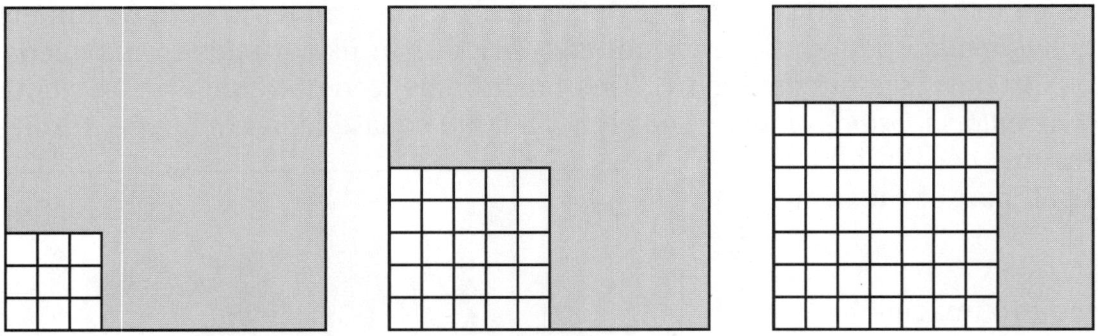

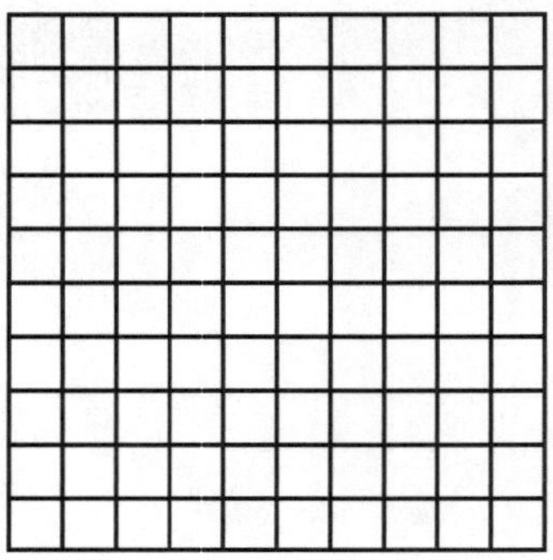

Make your own "growing white square" by sliding a piece of paper shaped like this one over the 10 × 10 square on the left.

2. **a.** How do you think the area of the white square will change as the length of a side increases from one to 10? Be specific.

b. Make a table to show the area of the white square for every length of a side from one to 10.

c. Make a graph that shows the relationship between side length and area using the information in your table.

d. Was your answer to part **a** correct? Explain why or why not using examples from your table.

3. What happens to the area of a square when the length of a side doubles?

You probably didn't find the area of each white square by counting all the small squares. You may have multiplied: 1 × 1 = 1, 2 × 2 = 4, 3 × 3 = 9, and so on. Multiplying a number by itself is called *squaring* a number.

4. Why do you think this term was chosen?

The numbers that result from squaring a whole number are called *square numbers,* or *perfect squares.*

5. Find at least five different square numbers. Share your list with a classmate. Each of you can try to guess what numbers the other person squared.

There are several ways to indicate the square of a number:

3^2, 3 ↑ 2, 3 times itself, and 3 ^ 2 all mean 3 × 3. Three squared is nine.

When using arrow language you can write:

$$14 \xrightarrow{\wedge 2} 196$$

Or, using words, you can write:

$$\text{length of side} \xrightarrow{\wedge 2} \text{area of square}$$

Recall that the *Armillaria bulbosa* fungus was found covering an area of 154,000 square meters.

Suppose that the fungus is in the shape of a square. To find the length of one side of the square, translate the problem to:

$$? \xrightarrow{\wedge 2} 154{,}000$$

The squaring has to be undone. In other words, you have to "unsquare" 154,000.

6. Estimate what number is the result of unsquaring 154,000. How can you check your estimation?

For some numbers, unsquaring is easy.

7. Write some numbers that are easy to unsquare. Then unsquare them.

8. Make up a situation in which you would have to unsquare 81.

$$? \xrightarrow{\wedge 2} 81$$

9. Do you think 576 is an easy number to unsquare? Explain.

10. Try to unsquare 150 by first making a guess. Then use your calculator to square your guess and see how close you are. Repeat this process two more times.

Normally, when arrow language is used, the number that is known is written at the left side of the arrow. If there is an arrow string in which the number on the left is missing, the string can be reversed so that the missing number is on the right. For example,

$$? \xrightarrow{+6} 20 \text{ becomes } 20 \xrightarrow{-6} ?.$$

It would be nice to be able to reverse the squaring arrow too. For example,

$$? \xrightarrow{\wedge 2} 81 \text{ would become } 81 \xrightarrow{?} ?.$$

But what symbol for unsquaring would go above the arrow?

Mathematicians have already invented a symbol that means unsquaring: $\sqrt{}$.
It is usually read as "taking the square root" instead of unsquaring.

11. Use the $\sqrt{}$ key on your calculator to find the square root of 81.

12. a. Write an arrow string showing the square root of 81.

You can also use the square root symbol without using arrows: $\sqrt{36} = 6$. (Notice that the number being unsquared is under the square root symbol.)

b. Use the square root symbol to write an equation that finds the square root of 81.

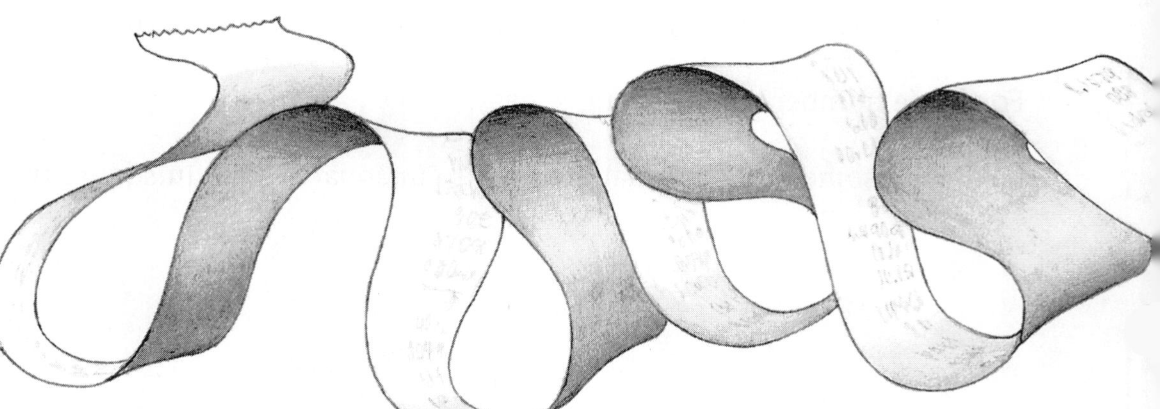

13. Find the square root of each of the following numbers. First make a guess and then use your calculator to check.

a. 25 **b.** 64 **c.** 121 **d.** $\frac{1}{4}$

14. a. Use the $\sqrt{\ }$ key on you calculator to find the square root of 150.

b. Compare the answer to the one you found in problem **10.**

c. What happens when you square the answer to $\sqrt{150}$? Did you expect that?

d. Write the answer your calculator gives you for $\sqrt{150}$ on a piece of paper. Clear the calculator and calculate the square of this number. If there is a difference, can you explain the difference between this number and the answer you got in part **a?**

For most numbers, it is not possible to find the exact square root because there are an infinite number of decimal places. So you never get the exact answer.

15. What does the calculator do since it cannot show a decimal that keeps going?

MORE HUMONGOUS FUNGUS

Remember the *Armillaria bulbosa* fungus that covers 154,000 square meters?

16. Find, to the nearest whole number, the length of each side of a square that would be covered by this fungus. Explain what you did to find it.

The actual site of the fungus is in Iron County, Michigan, near Crystal Falls.

17. a. On the map on **Student Activity Sheet 2,** draw a square (to scale) that represents the area of the fungus. (You may place the square anywhere on the map.)

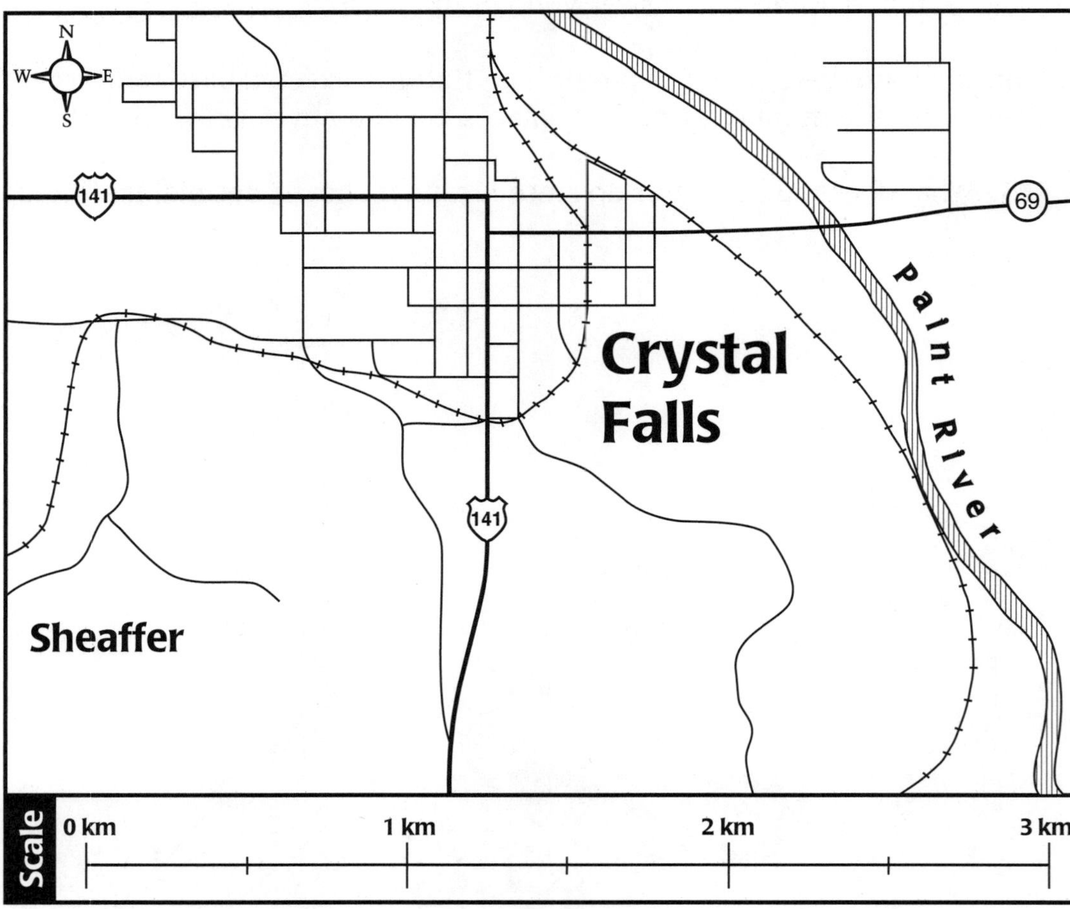

b. How did you decide on the length of the side for the square you drew?

Scientists think it is more likely that the fungus has the shape of a circle, rather than a square. If the shape of the fungus is a circle, you can find its area by using this formula:

$$\textbf{Area} = \pi r^2 \text{ or } \textbf{Area} = \pi \times \textbf{(radius)}^2$$

The symbol π (pi) represents a special number that describes the ratio of the circumference of a circle to twice that circle's radius. You may use 3.14 as an approximation for π. There also may be a π key on your calculator.

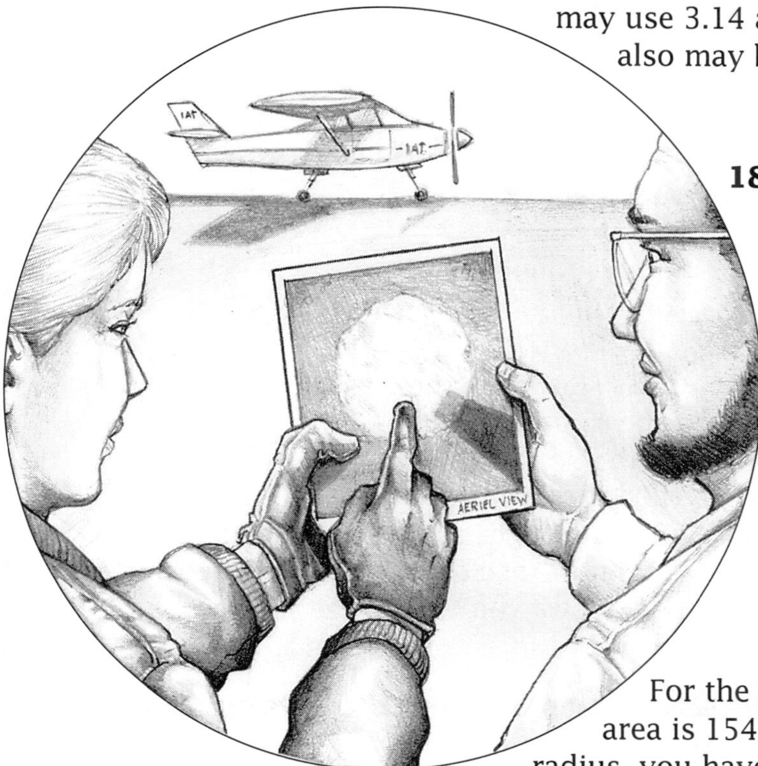

18. **a.** Make a table to show the change in the area of a circle as the length of the radius increases from 20 meters to 240 meters.

b. Make a graph that shows the relationship between radius and area.

c. How can you use your graph or table to find the radius of the 154,000-square meter fungus?

For the fungus problem, you know that the area is 154,000 square meters. To find the radius, you have to reverse the formula.

19. **a.** Write the area formula in arrow language and then reverse it so that you have a formula you can use to find the radius of a circle.

b. Use your reverse formula to find the radius of the circle covered by the fungus. Draw the circle on the map on **Student Activity Sheet 2.**

20. **a.** Compare the circle and the square that you also drew on the map.

b. Compare the graphs of the areas of squares and circles that you drew for problems **18b** and **2c.**

Summary

In this section, you studied perfect squares, square roots, and formulas for squares and circles.

If a number is a perfect square, then its square root is a whole number. For example, as shown on the right, 16 is a perfect square because $\sqrt{16} = 4$.

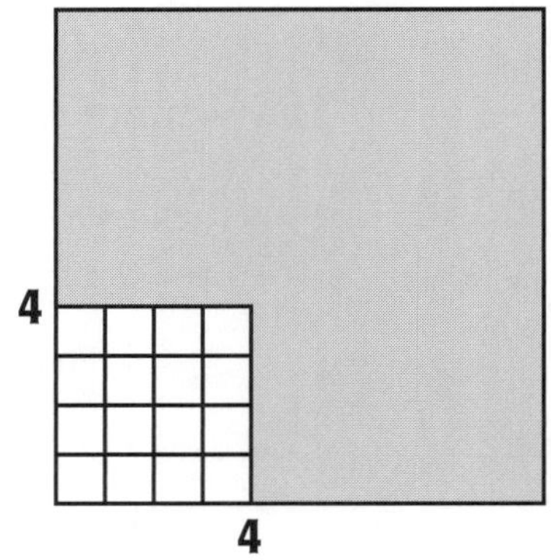

White Square = 16 Square Units

A number that is not a perfect square still has a square root, but its square root is not a whole number. For example, $\sqrt{1.44} = 1.2$, or, as shown on the left, $\sqrt{12}$ is approximately 3.464.

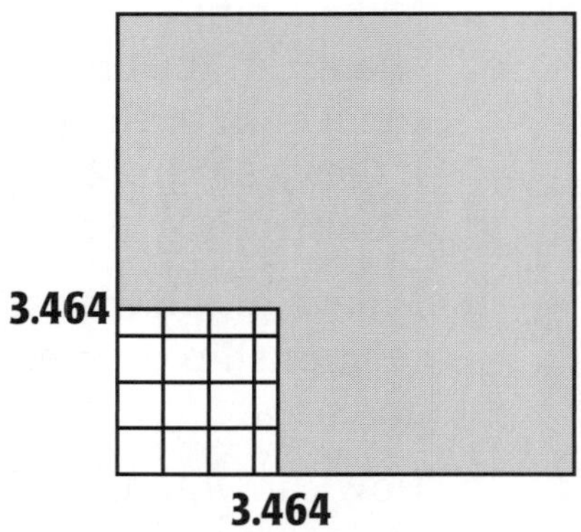

White Square = 12 Square Units

The unsquaring process allows you to find the length of a side of a square if you know the square's area. Unsquaring also allows you to find the radius of a circle whose area is known.

Summary Questions

21. **a.** Describe how to find the area of a square if you know the length of a side.

b. Describe how to find the length of a side of a square if you know its area.

22. List all of the numbers from 1 to 100 that are perfect squares.

23. Do you think that most whole numbers have whole numbers as square roots?

D. FORMULAS

Temperature

Dale and Kim are brother and sister. They have pen pals in Bolivia named Luís and Lucrecia, who are also brother and sister. Lucrecia and Luís are planning a visit to the United States, and they will stay with Dale and Kim's family.

Below you see the letter Lucrecia sent from Bolivia.

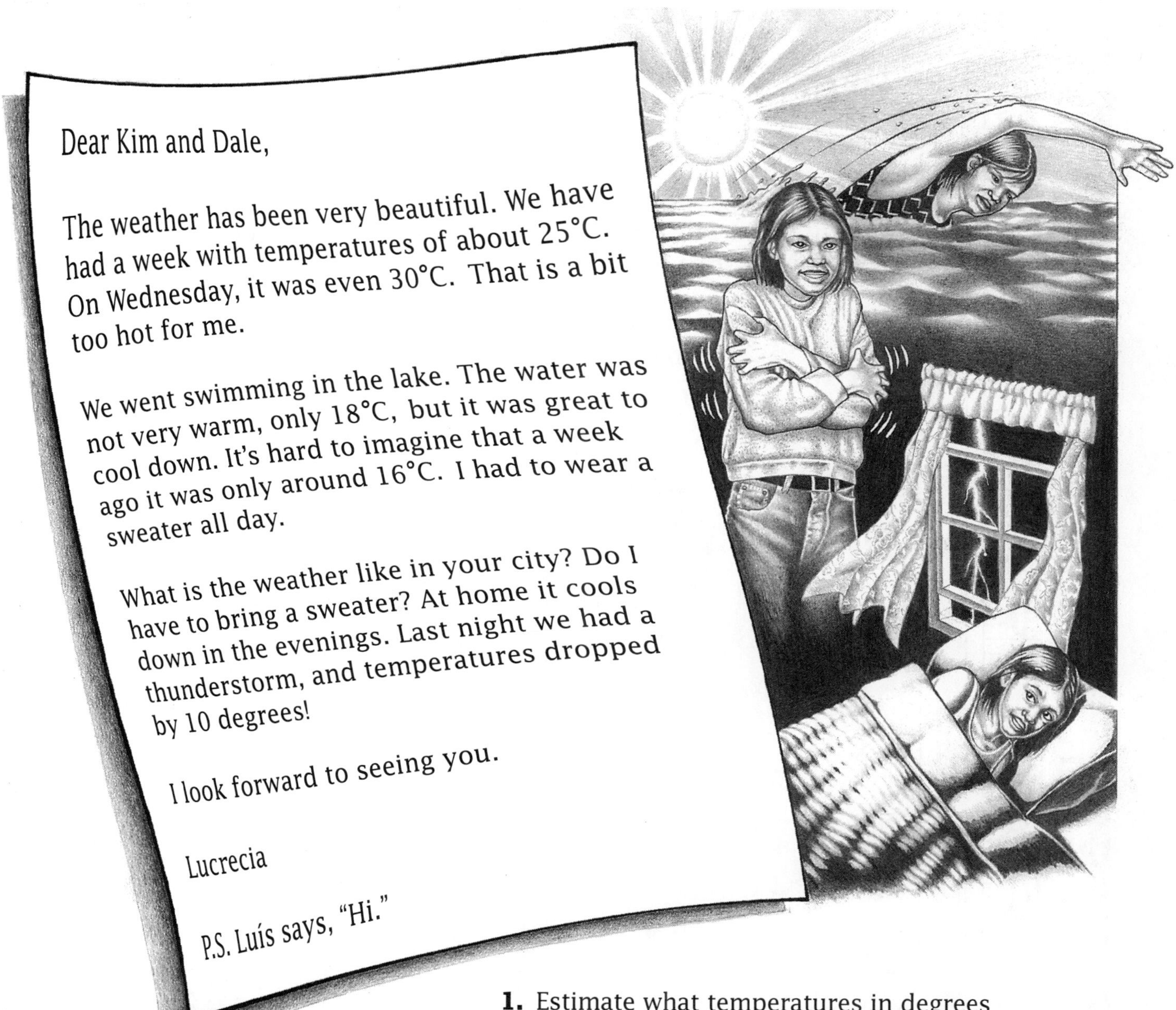

Dear Kim and Dale,

The weather has been very beautiful. We have had a week with temperatures of about 25°C. On Wednesday, it was even 30°C. That is a bit too hot for me.

We went swimming in the lake. The water was not very warm, only 18°C, but it was great to cool down. It's hard to imagine that a week ago it was only around 16°C. I had to wear a sweater all day.

What is the weather like in your city? Do I have to bring a sweater? At home it cools down in the evenings. Last night we had a thunderstorm, and temperatures dropped by 10 degrees!

I look forward to seeing you.

Lucrecia

P.S. Luís says, "Hi."

1. Estimate what temperatures in degrees Fahrenheit are the same as the Celsius temperatures Lucrecia mentions in her letter.

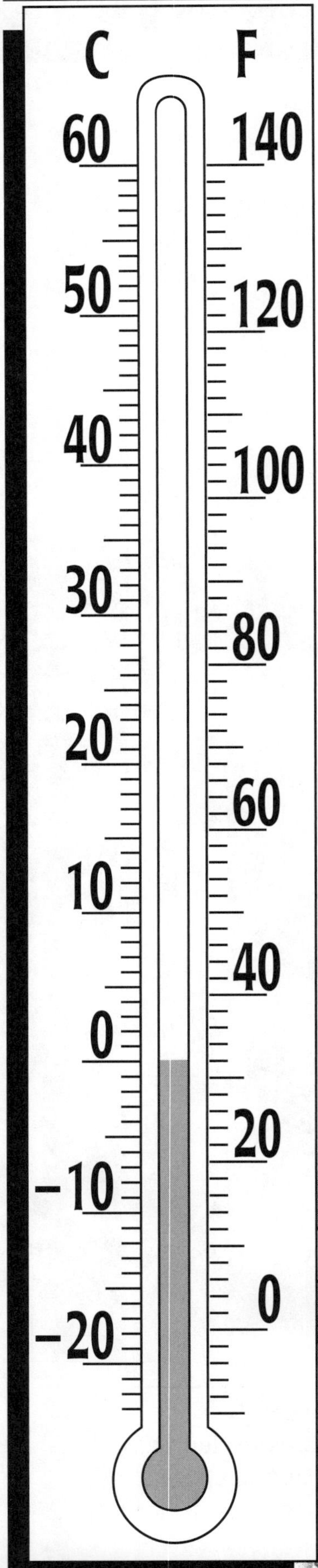

On the left is a thermometer that shows both Fahrenheit and Celsius temperatures.

2. a. How can you use this thermometer to find the answers to problem **1?**

b. Check the estimates you made for problem **1.** Were they close to what the thermometer tells you?

3. a. Look carefully at the thermometer. An increase of 10°C corresponds to an increase of how many °F?

b. Use your answer to part **a** to answer the following question. An increase of 1°C corresponds to an increase of how many °F?

c. Could you have answered part **b** just by looking at the thermometer? Explain.

When you use a thermometer to convert temperatures, you sometimes have to estimate the degrees because of the way the scale lines are drawn.

4. Do you think it is possible to calculate a Fahrenheit temperature for each Celsius temperature? Why or why not?

Here is a formula for converting temperatures from degrees Celsius (C) into degrees Fahrenheit (F).

$$\mathbf{1.8 \times C + 32 = F} \text{ or } \mathbf{1.8C + 32 = F}$$

5. a. Explain where the numbers in the formula come from. (*Hint:* Use the thermometer and your answer for problem **3b.**)

b. Write the formula using arrow language.

The formula and the thermometer tell you the relationship between C (the temperature in degrees Celsius) and F (the temperature in degrees Fahrenheit).

You can also make a graph to show the relationship.

C	−20	−15	−10	−5	0	5		
F								

6. a. First, fill in the table on the top of **Student Activity Sheet 3.** (Add some temperatures of your own choice, too.)

b. Describe any patterns you see in the table.

c. Graph the information from the table on top of **Student Activity Sheet 3.** (Notice that C is on the horizontal axis and F is on the vertical axis.)

d. Describe the graph.

7. How could you tell that the graph would be a line?

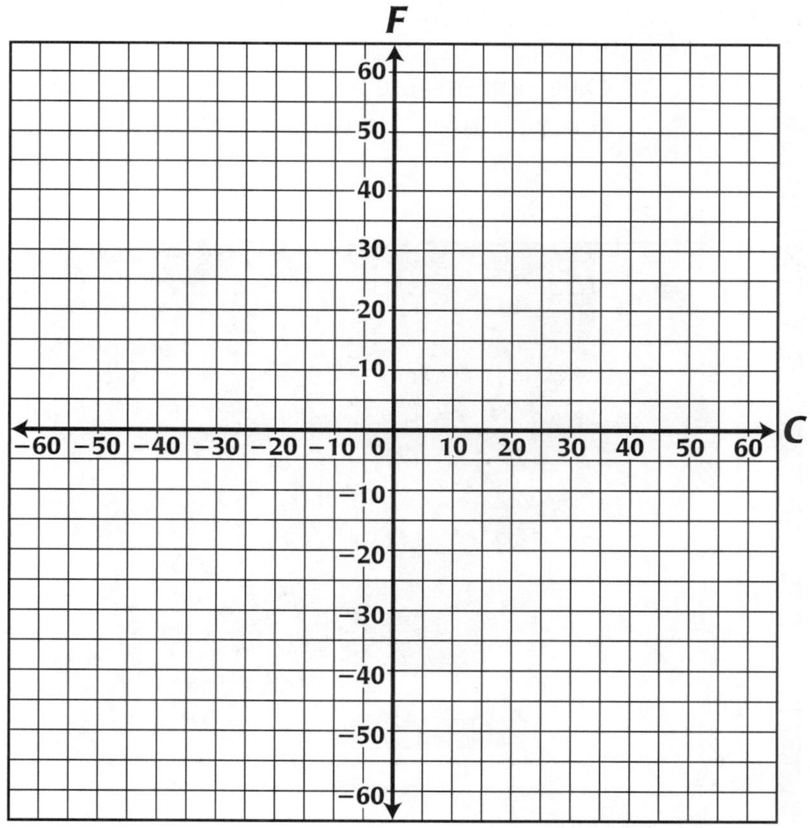

8. There is only one temperature that has the same value in degrees Celsius and degrees Fahrenheit. What temperature is this? Describe how you found your answer.

To convert temperatures, you can use a thermometer, a graph, a table, or a formula. There are many ways to write a formula that converts between Fahrenheit and Celsius.

9. Write a formula that converts between Fahrenheit and Celsius other than the one given on page 37.

Most formulas for converting between the two types of degrees are not easy to use if you are trying to make mental calculations. Here is a method you can use when converting mentally. To change Celsius to Fahrenheit, do the following:

Double the Celsius value, subtract $\frac{1}{10}$ of the result, and add 32.

10. Convert four temperatures mentally using this procedure.

11. **a.** Write the procedure as a formula.

b. Compare your formula with the formula given on page 37. What similarities and differences do you see?

Dale remembered a formula he learned in school for converting temperatures. He used an arrow string to write it on the chalkboard.

Kim wondered if this could be a correct formula. She said, "First you add 40, and then you subtract 40, so nothing happens. You can skip those two arrows."

12. Do you agree with Kim? Why or why not?

Here are some pictures of Kim, Dale, Luís, and Lucrecia at various places.

13. What is a likely temperature in degrees Celsius for each picture?

While Lucrecia and Luís are in the United States, they want to convert from Fahrenheit to Celsius so that they can understand what the temperature is. They hear on the radio that the high for the day will be 30°F.

14. Describe how you can find what 30°F is in degrees Celsius.

15. Write a formula that you can use to convert temperatures from *F* into *C*. (*Hint:* You may be able to adapt one of the formulas you have been working with.)

Heavy Training

Your heart rate when you are lying or sitting is considered your normal heart rate. When you fall asleep your heart rate slows, and when you exercise or are upset your heart rate increases.

Activity

16. With a partner, find your normal heart rate (*N*). To do this, find your pulse in your neck or wrist and count the beats for 20 seconds. Your partner should watch the clock and tell you when to start counting and when to stop.

Heart rate is usually reported in terms of beats per minute. Use the pulse that you counted in 20 seconds to find your normal heart rate in beats per minute.

Switch roles with your partner and repeat the above procedure so that both of you know your normal heart rates.

Athletes who take part in endurance sports need to be in very good condition. When they compete, their heart rates increase.

Because it is dangerous for a person's heart rate to be too high for too long, athletes train specifically to increase their endurance. It is important for athletes to determine their maximum heart rate (M).

Finding the exact value of an athlete's maximum heart rate is difficult. There is, however, a formula that gives a close approximation.

$M = 220 - A$, where A is the age of the athlete in years

17. Make a graph on **Student Activity Sheet 4** that could be placed in a gym to help people find their maximum heart rates.

Use the above formula and the graph you made for problem **17** to answer the following questions.

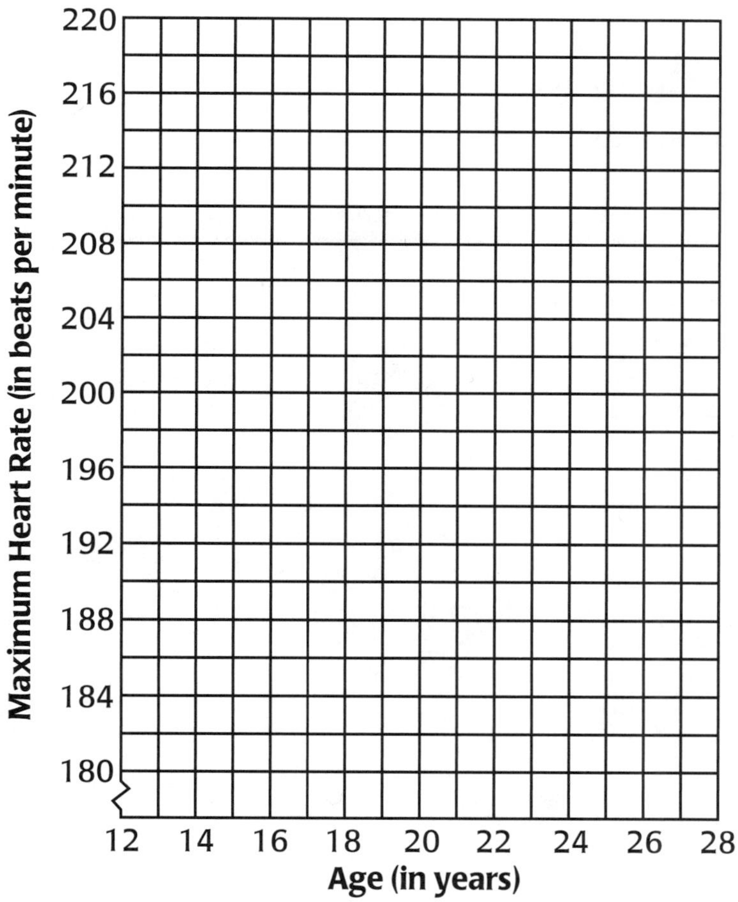

18. a. Who has a higher maximum heart rate: you or your teacher? Explain.

b. What can you say about the relationship between age and maximum heart rate?

c. What is the highest possible maximum heart rate? the lowest?

One training method depends on something called the *reserve heart rate*. To calculate the reserve heart rate (R), you take the difference between the maximum heart rate (M) and the normal heart rate (N) of the athlete.

19. a. Write a formula for finding the reserve heart rate.

b. Why do you think it is called the reserve rate?

c. What do you need to know in order to calculate a person's reserve rate (R)?

20. Find your reserve heart rate. Explain the calculations you made.

Warning: The M value can vary with your physical condition. You should not use the above formula to gauge your own workouts without consulting a physician.

Now you have two formulas: $M = 220 - A$ and the one for R that you found for problem **19.** These can be rewritten as tree diagrams.

Here is a tree diagram for the formula for M.

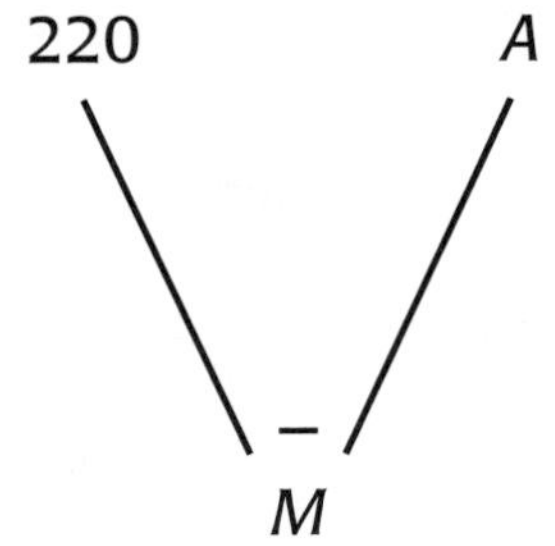

21. **a.** Make a tree diagram for the formula for R.

b. Combine the tree diagrams to make one that can be used to calculate R.

c. Write one formula to fit this tree.

d. What values do you need to know to calculate R?

22. Calculate the reserve heart rate (R) for these athletes.

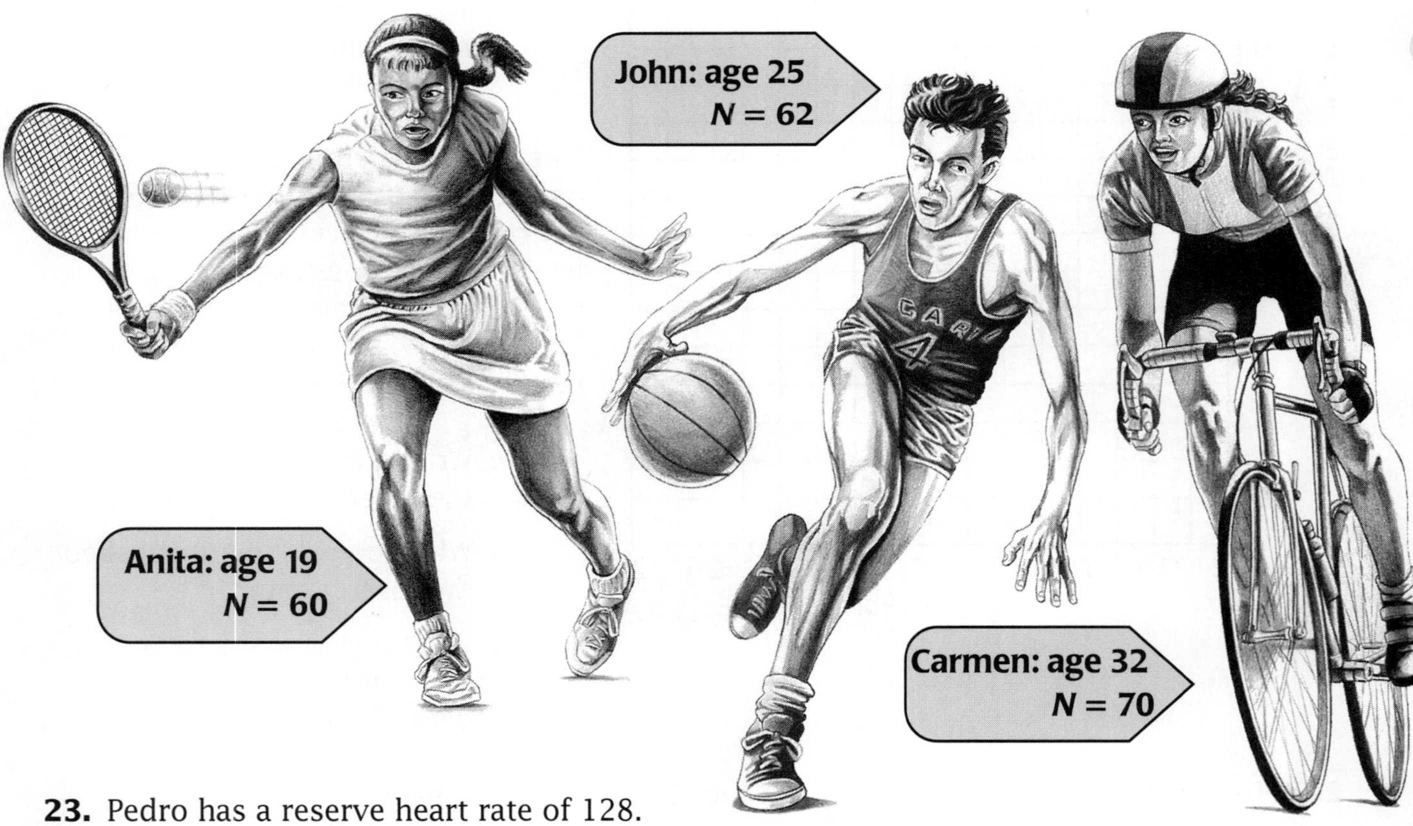

23. Pedro has a reserve heart rate of 128.
What can you determine about his age and normal heart rate?

During a training session, not all of the reserve heart rate is used. A trainer develops a training program that allows each athlete to use a certain amount of his or her reserve heart rate.

Jacob is the trainer for John, Anita, and Carmen. He decides that these athletes should be using 75% of their reserve heart rates at the peak of their workouts. This optimal heart rate is called the training value (T).

24. **a.** Calculate each of the athletes' training values.

b. If you were an athlete training with Jacob, what would your training value be?

In a large training camp, these calculations are performed by a computer. The trainer or athlete types in the values for age (A) and normal heart rate (N) and the percentage of reserve heart rate (R) that will be used. The computer then displays the training value (T).

25. Design a tree diagram and write a formula for T.

Building Stairs

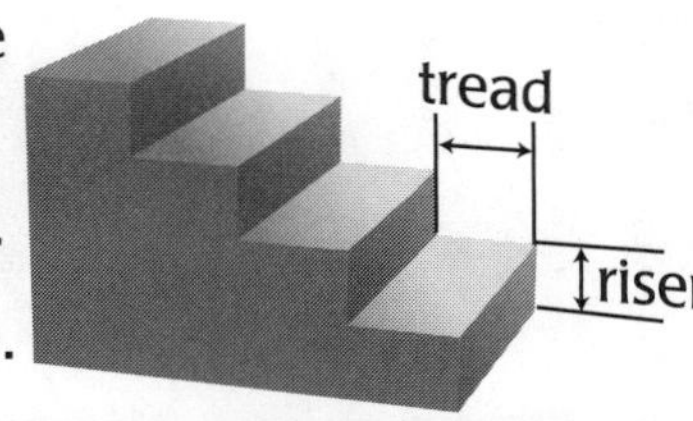

Here you see a picture of a staircase. All of the steps are the same size. Each step has two main parts called the *riser* and the *tread.*

The vertical measure, or the height, of a step is called the *rise* (R).

The horizontal measure, or depth, of a step is called the *tread* (T).

Activity

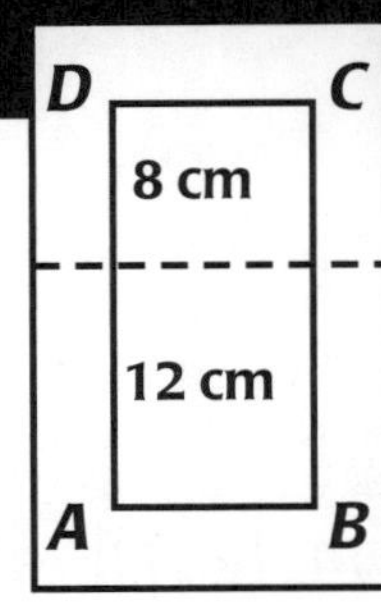

You are going to build a model staircase like the one shown above out of stiff paper.

i. Draw a rectangle that is exactly 20 centimeters long and 10 centimeters wide in the center of a piece of stiff paper. Label the corners of the rectangle *A, B, C,* and *D* as shown in the diagram above.

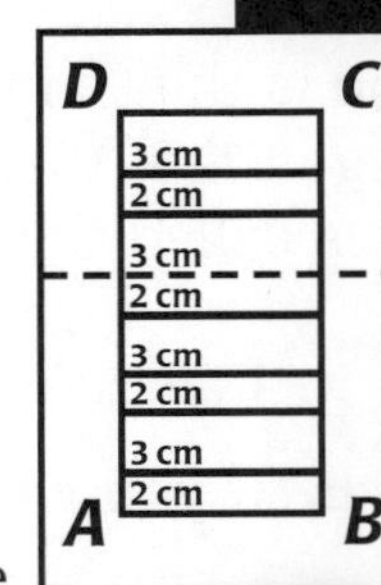

ii. Draw a dotted line across your paper that is 8 centimeters below the top of the rectangle as shown above.

iii. Fill the rectangle with lines that are alternately 3 and 2 centimeters apart, as shown in the diagram on the right. (It is easy to keep your lines parallel with a ruler and a triangle as the girl in the picture on the left is doing.)

iv. Fold the paper along the dotted line and cut along the long sides of the rectangle. Do not cut along the short sides.

v. Fold the solid lines like an accordion so that you end up with a staircase. The first fold should be on side *DC* folded toward you (out). Fold the next line away from you (in). Continue alternating the fold direction until the staircase is finished.

You now have a model staircase.

vi. Label the wall and the floor on your model as shown on the right.

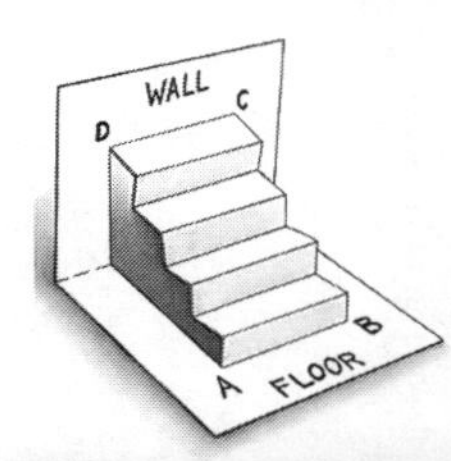

The stairs you made fit nicely with the floor and the wall, in other words, the treads and the floor are perfectly horizontal and the risers and the wall are perfectly vertical.

26. Do you think this is a coincidence or were they designed that way? Why do you think so?

27. a. Measure and record the height and depth of the whole staircase (depth is measured along the floor).

b. What are the values for *T* and *R* in the steps you made?

c. How are the height and depth of the whole staircase related to the rise and tread of each step? Explain.

28. On your model staircase, make the fold between the floor and the wall in a different place. Is the tread of each step in your model still perfectly horizontal? Is the rise of each step exactly vertical? Explain why or why not.

29. a. Would the designs shown below make good staircases? Explain.

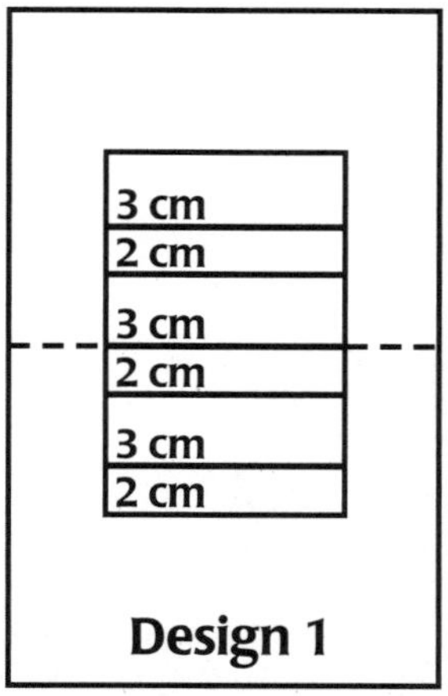

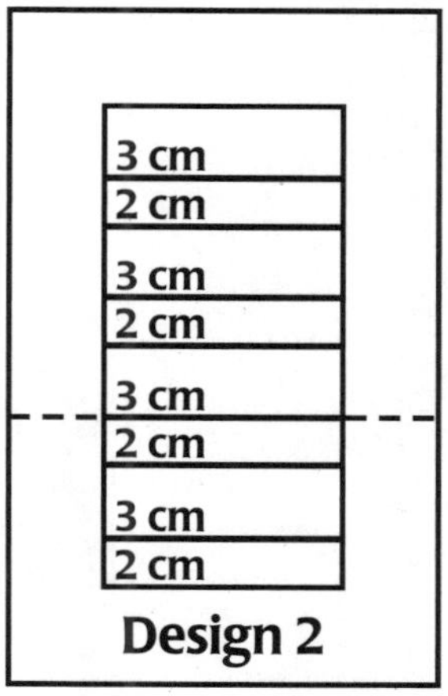

b. Copy the drawing below into your notebook. Draw a fold line where it will create a good model staircase.

c. What are some rules for making good model staircases?

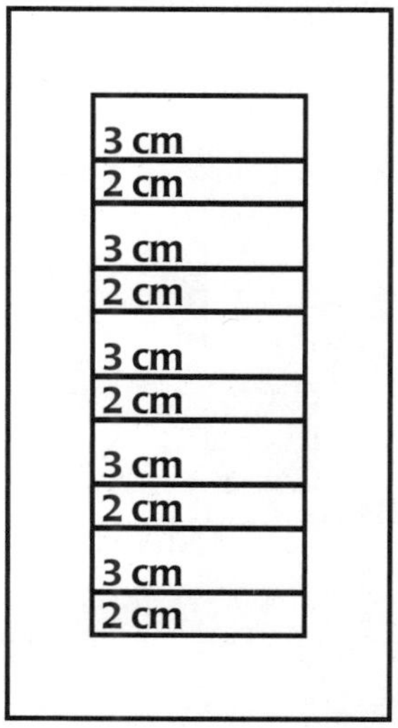

Not all stairs are easy to climb.

30. **a.** Write down some things that influence how easy it is to climb a set of stairs.

b. Order the staircases shown on this page according to how easy you think they would be to climb. Give reasons to support your choices.

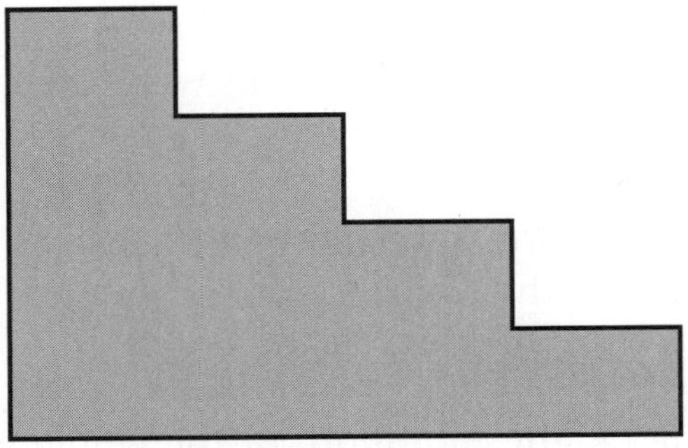

Staircase A

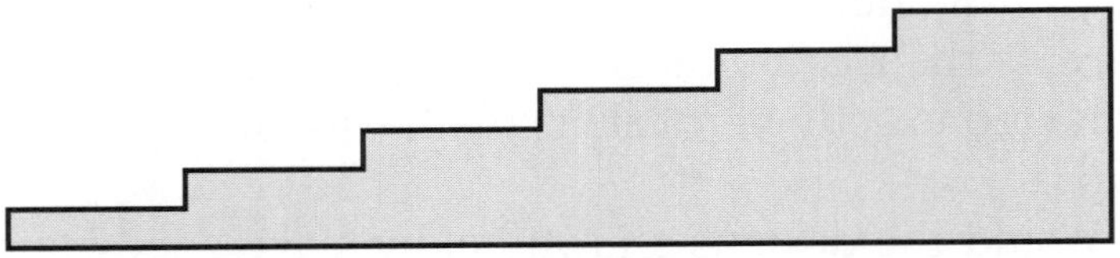

Staircase B

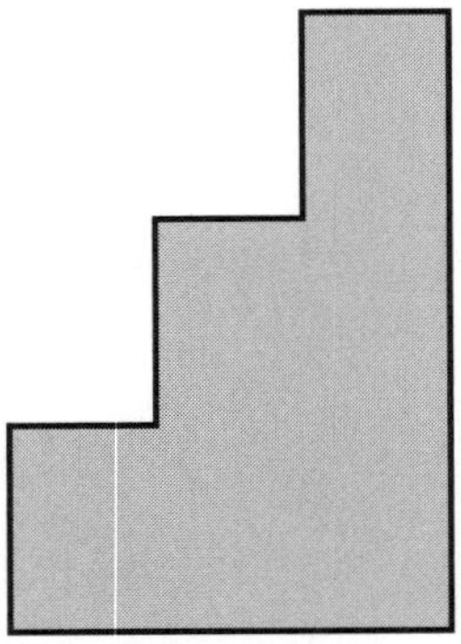

Staircase C

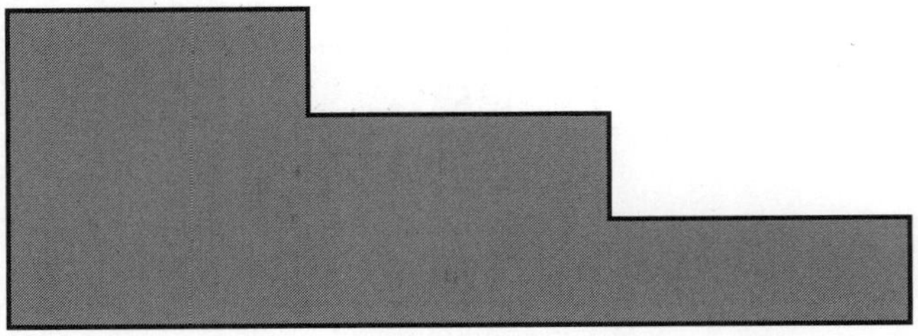

Staircase D

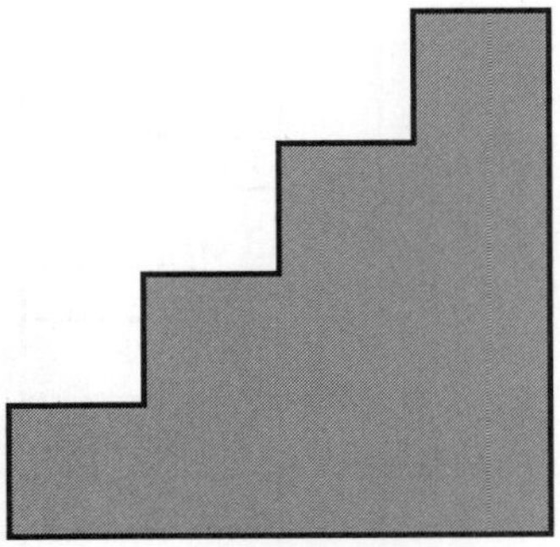

Staircase E

Activity

Measuring Stairs

Measure a flight of stairs that you use regularly at school, home, or somewhere else. Count the steps. Measure the risers and treads. Be sure to pay attention to how the steps are made.

They could be like this:

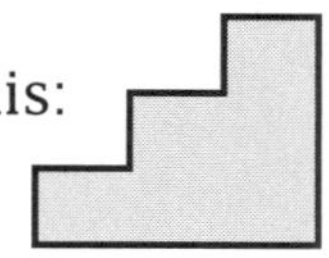

or like this:

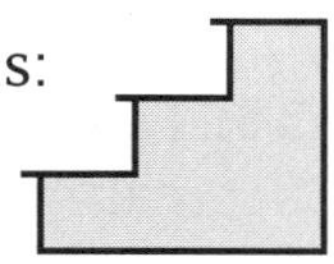

If your steps are of the second type, measure the tread in two parts as shown.

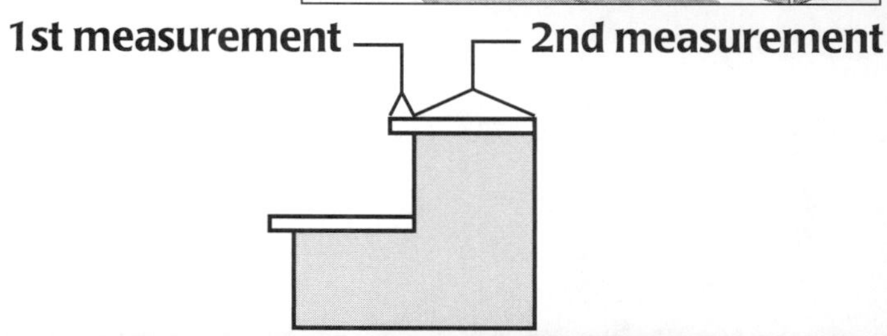

Name	Stair Location	Kind of Step	Rise (in cm)	Tread (in cm)	Number of Steps	Easy to Climb?
Monica	office building		18.5	3.5 + 23.5	9	yes

31. a. Record the measurements in a table like the one above.

b. Make a scale drawing of the steps. (Be sure to include the scale.)

c. Calculate the height and depth of the whole flight of stairs.

32. Compare the stairs that were measured by members of the class. Did all of the class members use the same scale for their drawings?

Some flights of stairs are easier to climb than others. The steepness of the stairs is one factor that affects how easy they are to climb.

33. **a.** Look at the drawing you made of the stairs for problem **31.** Draw a new staircase that has the same total height as the one you drew, but that is steeper.

b. What did you have to do to make the stairs steeper?

c. What are some advantages and disadvantages of steep stairs?

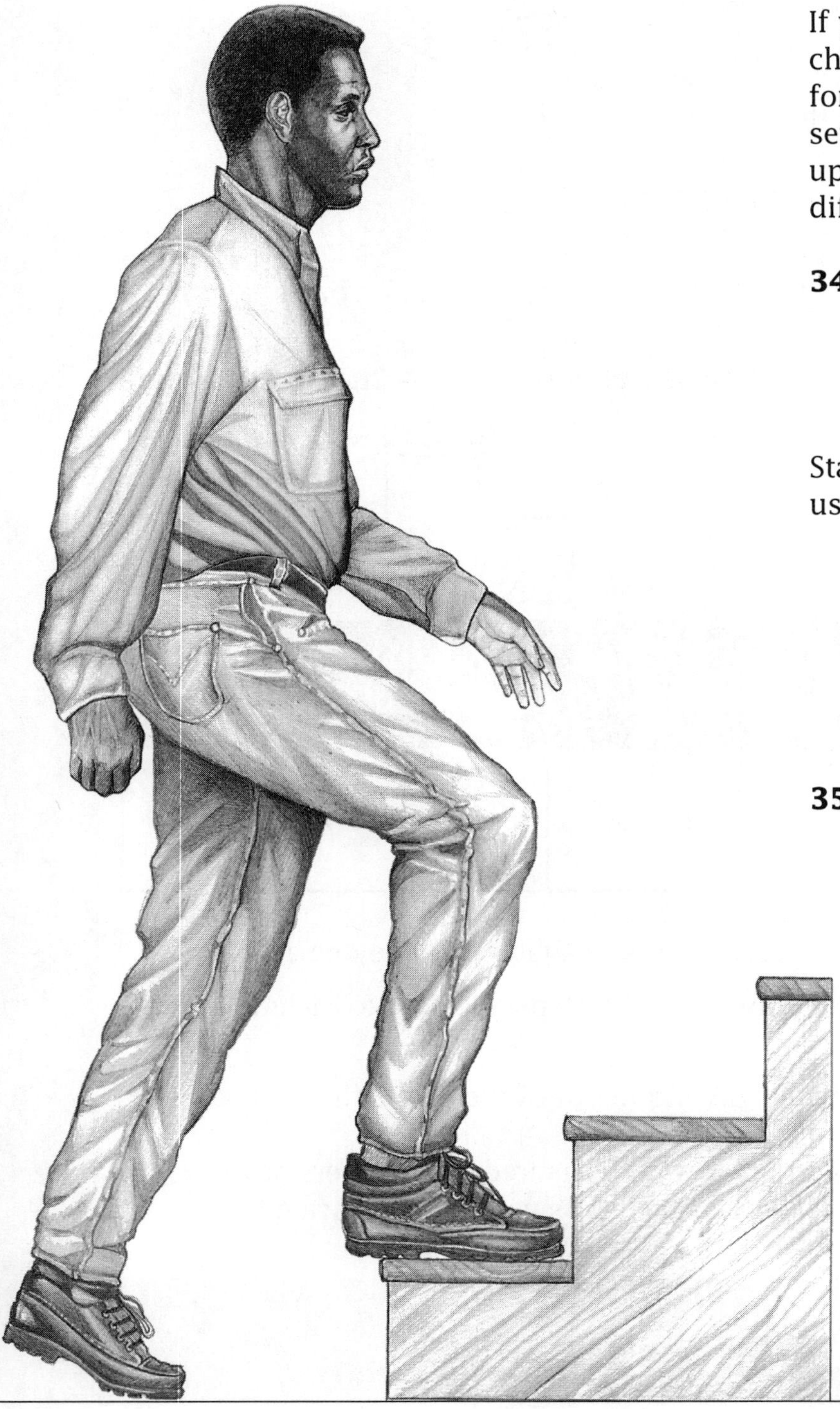

If you are not careful about choosing the measurements for the rise and tread of a set of stairs, you can end up with stairs that are difficult to climb.

34. List five things that can make a set of stairs difficult to climb. Be specific.

Stairs that are easy to climb usually fit the following rule:

$$2 \times \text{Rise} + \text{Tread} = \text{Length of One Pace}$$

or

$$2R + T = P$$

35. **a.** Does this rule seem reasonable to you? Why or why not?

b. Use the information gathered from the class in problem **31** to calculate $2R + T$ for all of the stairs that were measured.

c. Compare the values you found. What does this tell you about how stairs are built?

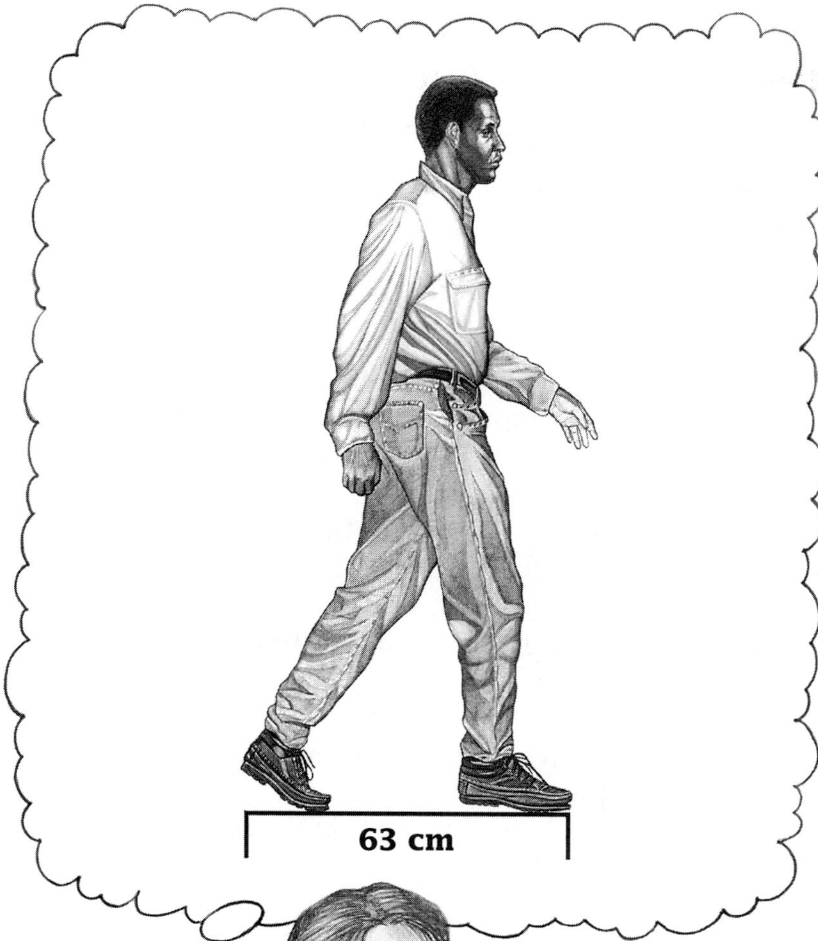

An adult's pace is about 63 centimeters. So the rule can be written as follows:

$$2R + T = 63$$

36. **a.** A contractor wants to build a set of stairs with a rise of 19 centimeters for each step. What size will the tread be if she follows the above rule? Explain.

b. For another set of stairs, the contractor knows that the tread must be 23 centimeters. How high will each rise be if the contractor uses the rule? Explain.

You have now found two combinations of rise and tread measurements that fit the rule based on an adult pace of 63 centimeters.

37. a. Find a few more pairs of numbers that fit the rule.

b. On **Student Activity Sheet 5,** graph all of the pairs of rise and tread measurements that fit the rule.

You can make stairs that are difficult to climb even when you use the formula.

38. Which points on the graph would represent stairs that are difficult to climb?

39. What happens to the tread (T) if you add 1 centimeter to the rise (R) and are using the rule? Can you see this on your graph?

40. Using the rule, when do R and T have the same value?

41. Which of the stairs that were measured by members of the class for problem **31** fit the rule exactly? Which come close?

Here is another rule that helps in designing stairs that are easy to climb:

$$\text{Rise} \leq 20 \text{ centimeters}$$

This means that the rise is less than or equal to 20 centimeters.

42. a. Why would this rule make stairs easier to climb?

b. Find a way to show this rule on your graph.

c. Find measurements for some stairs that fit both rules.

d. There are some situations that do not allow for stairs that are easy to climb. What could be some reasons for bad stairs?

Think about the dimensions of the paper stairs you made earlier. Suppose the paper stairs are a model for an actual set of stairs that uses the rule $2R + T = 63$ centimeters.

43. a. What are the measurements for the rise and tread in the actual flight of stairs?

b. What are the height and depth measurements of the whole flight of stairs?

Here are some other rules used for building stairs in different kinds of buildings.

Private Homes
Rise – maximum 20 cm
Tread – minimum 23 cm

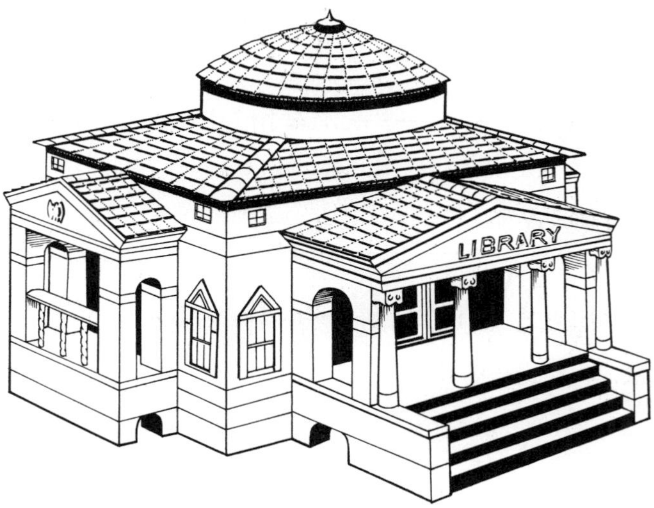

Public Buildings
Rise – maximum 18 cm
Tread – minimum 28 cm

44. **a.** Why do you think that there is a maximum for the rise?

b. Why is there a minimum for the tread and not a maximum?

45. Design a staircase for a public building with a total height of 3 meters. The staircase should take up as little floor space as possible (it should have the smallest possible depth measurement). Make sure it fits the rule above for a public building and the rule $2R + T = 63$ centimeters.

Egyptian Art

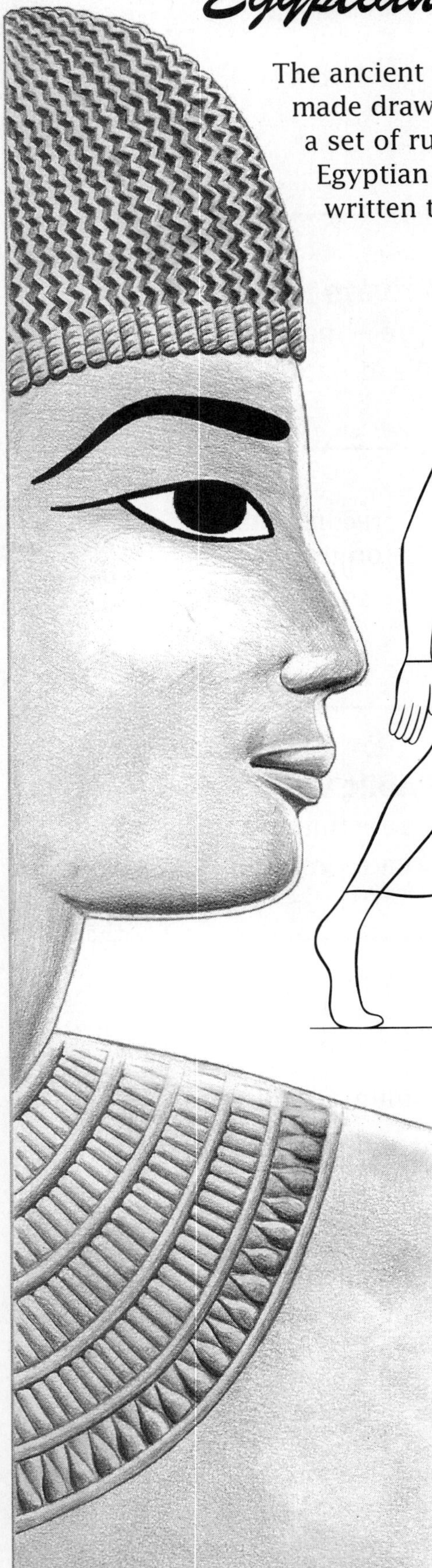

The ancient Egyptians were fascinated by proportions. When they made drawings or sculptures of people, the measurements followed a set of rules. These rules help modern archaeologists reconstruct Egyptian pictures that have been damaged. Archaeologists have written these rules as formulas that are easy to use.

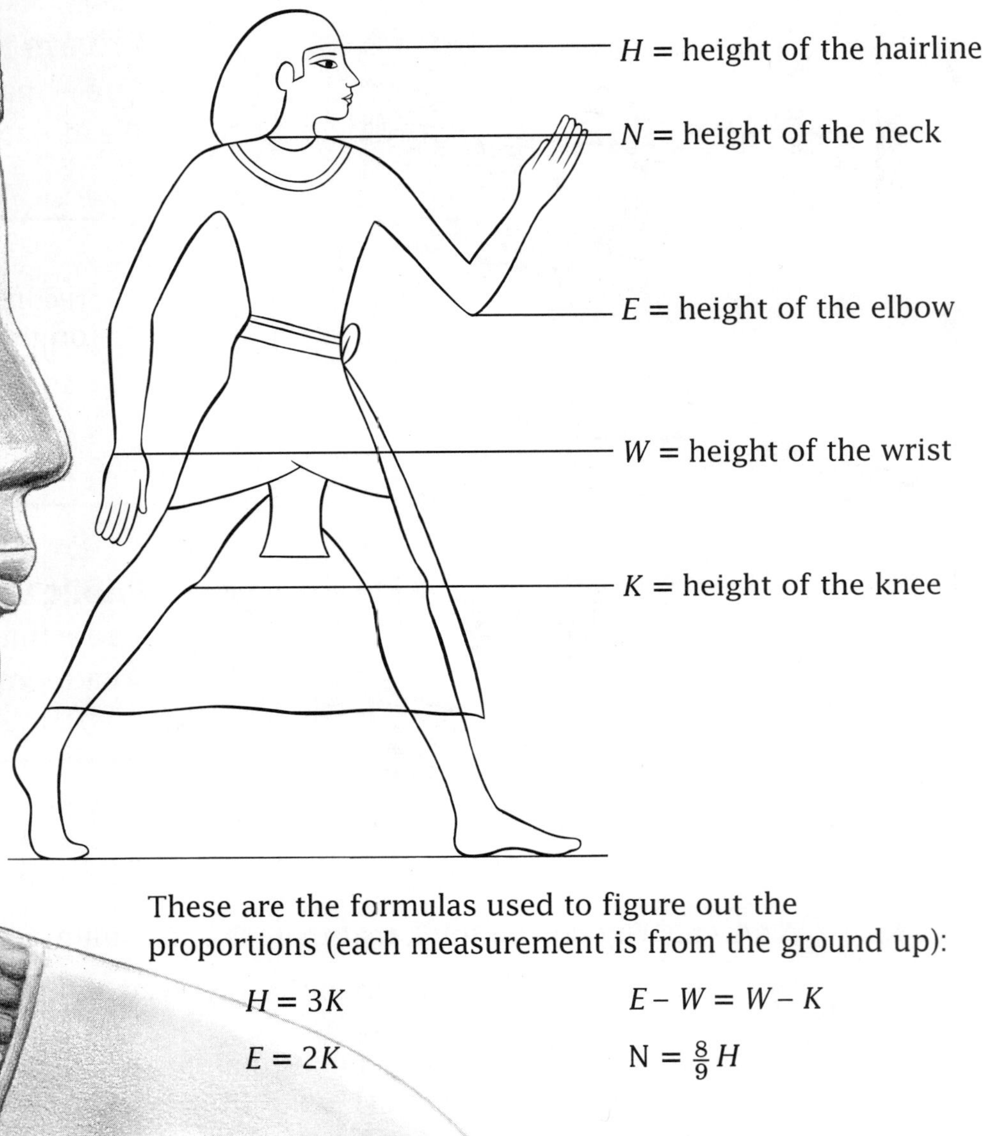

These are the formulas used to figure out the proportions (each measurement is from the ground up):

$H = 3K$ $\qquad E - W = W - K$

$E = 2K$ $\qquad N = \frac{8}{9}H$

46. **a.** Translate the formula $E = 2K$ into an English sentence about the drawing.

b. Describe how you can check whether or not $E - W$ equals $W - K$ in the drawing above.

c. Do the measurements of the Egyptian in the picture above fit the given formulas?

Activity

On a recent excavation, archaeologists discovered pieces of an ancient Egyptian drawing. Unfortunately, parts of the drawing were missing. The archaeologists made careful tracings of the pieces they found.

47. Use **Student Activity Sheet 6** to reconstruct the original drawing.

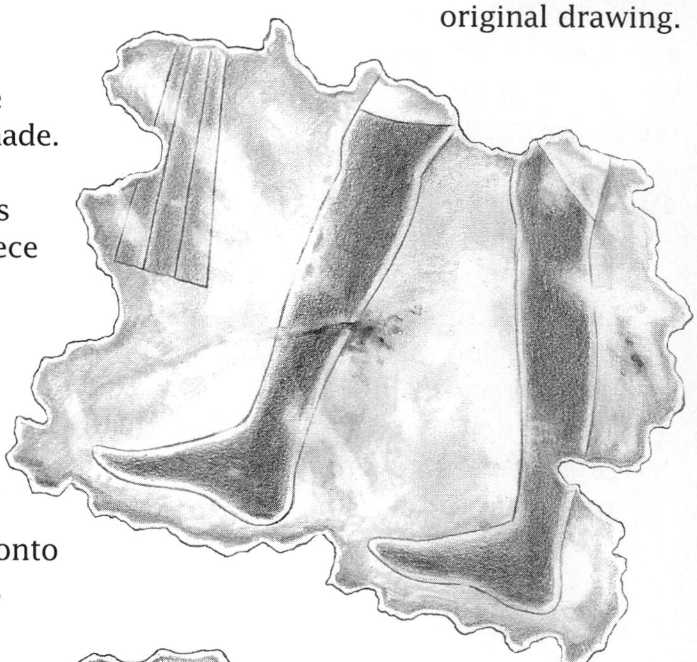

i. First cut out the tracings that the archaeologists made.

ii. Use the formulas to place each piece correctly. (Keep track of your calculations so that you can explain your process to someone else.) Glue the pieces onto a sheet of paper.

iii. Then sketch the missing portions.

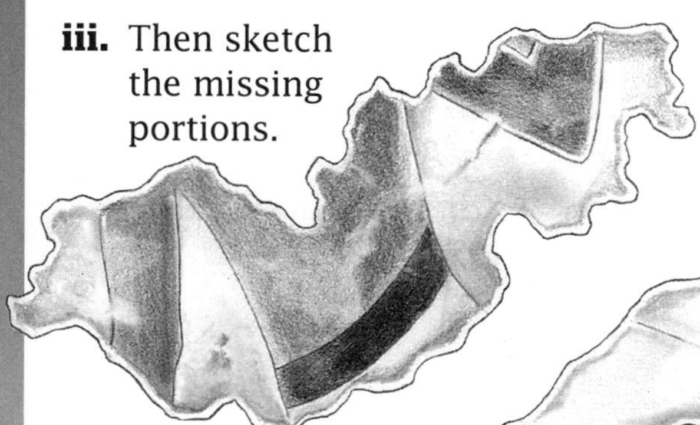

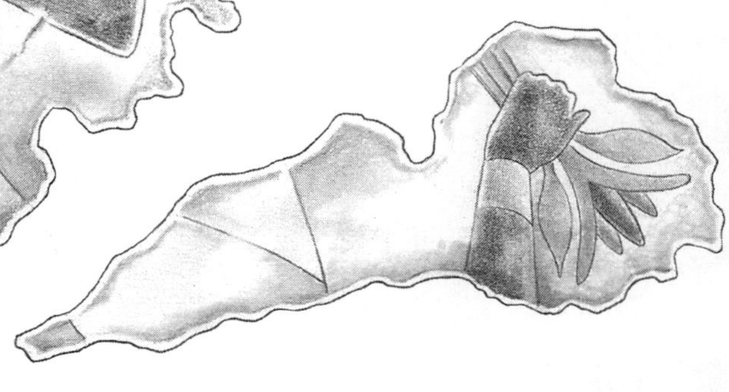

Later in the dig, the team of archaeologists found fragments of another drawing. This drawing had grid lines over it, which the ancient Egyptians may have used to check the proportions. The archaeologists thought that this drawing would be easier to reconstruct than the first one because of the lines. After they looked at the pieces, however, they realized that it would be harder.

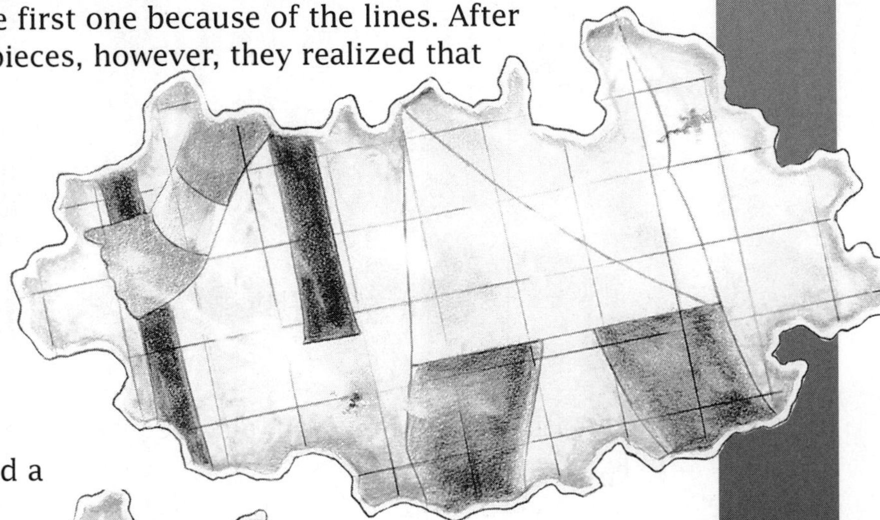

48. Why is this drawing harder to reconstruct?

After looking carefully at the formulas, the archaeologists found a way to get around the problem.

49. Use **Student Activity Sheet 7** to reconstruct this picture. Cut out the pieces, glue them into the correct positions, and sketch the missing parts. (Remember to keep track of your calculations.)

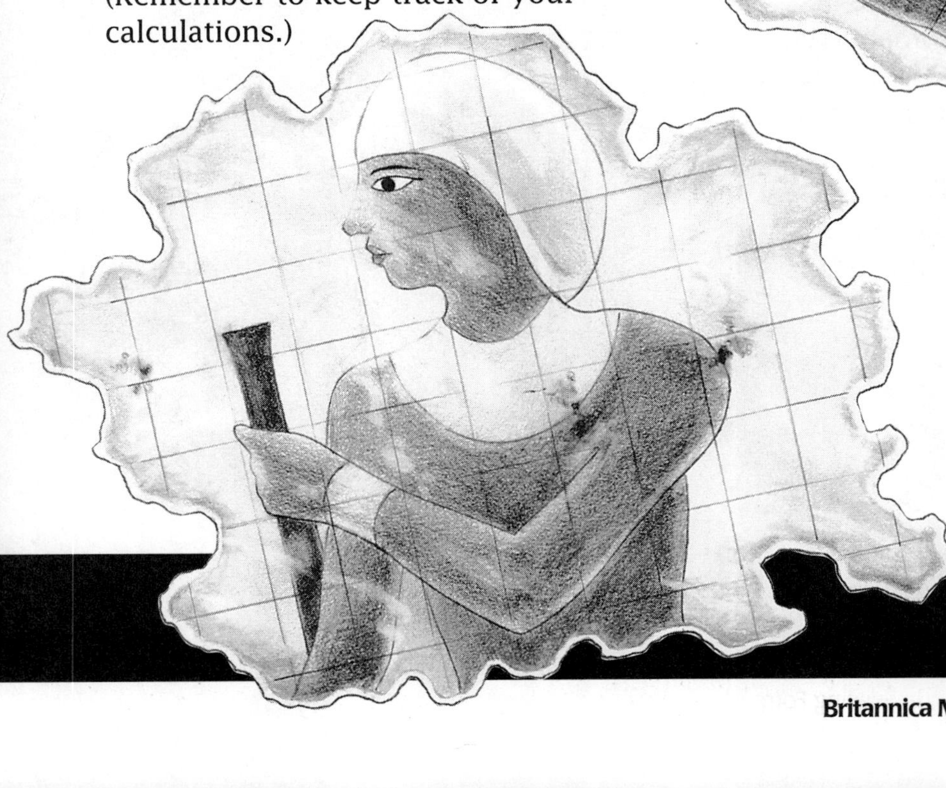

Summary

In this section, you have seen how formulas are used in a variety of situations. Formulas can be used to:

- describe a situation, as was the case with the Egyptian art, for which formulas describe how drawings were made;
- solve problems, such as how to convert temperatures;
- investigate possibilities within certain constraints, such as how to build a staircase that has a total height of 3 meters but is easy to climb.

You will encounter many more situations in which formulas can be used.

Information in formulas can also be given in pictures, stories, graphs, tables, or other formulas.

When solving problems, be sure to choose the best way to organize and represent the information.

Summary Questions

Do not believe everything you read! The following was printed on the cover of a notebook:

- To convert Fahrenheit temperatures to Celsius temperatures, use this formula: $C = \frac{5}{9}(F - 32)$.
- To convert Celsius temperatures to Fahrenheit temperatures, use this formula: $F = \frac{9}{5}(C + 32)$.

50. Are these formulas correct? Write a letter to the company that produced the notebook to explain why or why not.

Summary Questions, continued

A rule for building an exit ramp states that the vertical distance of the ramp must be no more than one-eighth the horizontal distance.

51. Which of the following ramps fits the above rule?

a.

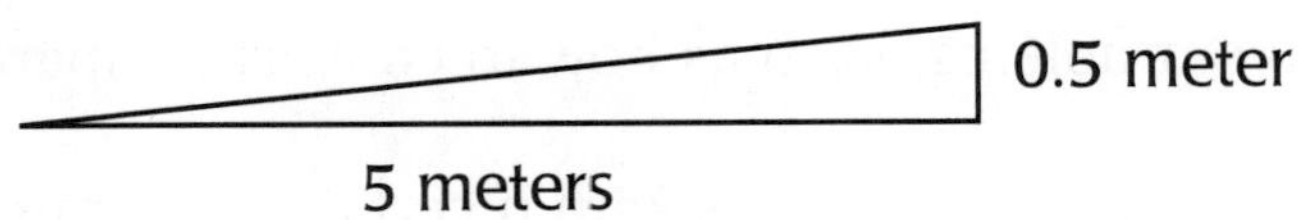

b.

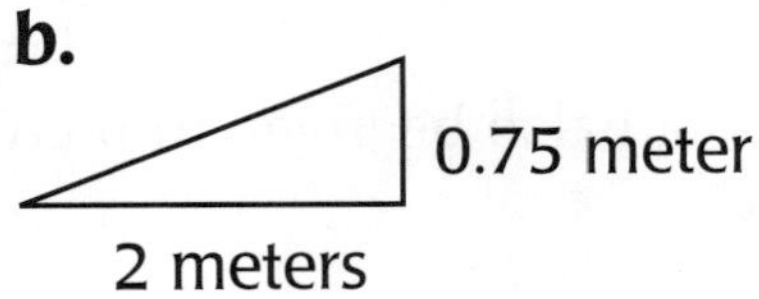

c.

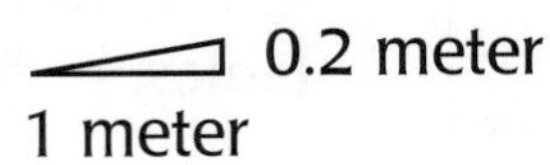

52. Write the above rule for an exit ramp as a formula.

One of the factors that affects how long it takes a driver to stop a car is the speed of the car. Suppose the following formula finds the stopping distance in feet if you know the speed of the car.

$$7 \times \text{Speed} - 74 = \text{Stopping Distance}$$

53. What is the stopping distance if the car's speed is 20 miles per hour? 40 miles per hour? 60 miles per hour?

54. **a.** Create a graph of the speed and stopping distance situation. Place the speed along the horizontal axis and the distance along the vertical axis.

b. Are there any restrictions for possible speed and/or distance values? Explain.

TRY THIS!

Section A. Patterns

On the right is a drawing of a path. The path uses gray and white hexagonal tiles. Tiles with a hexagonal shape have six equal sides.

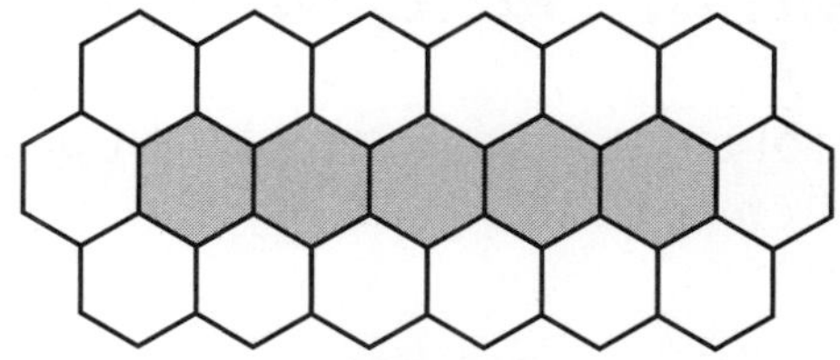

Length = 7

The length of a path is the number of gray and white tiles from left to right. In the above example, the length is seven. The path drawn on the right has a length of eight.

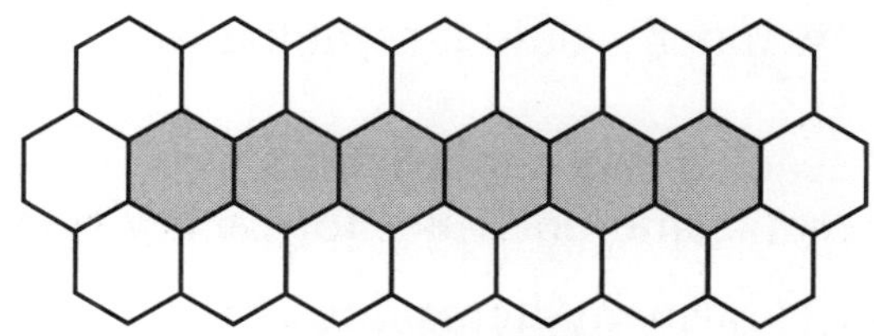

Length = 8

1. Suppose you want to make a path with a length of nine.

a. How many white tiles do you need?

b. How many gray tiles?

c. How many tiles in total?

2. Write a direct formula to:

a. describe the number of white tiles when you know the length of a path.

b. describe the number of gray tiles when you know the length of a path.

3. a. Now, write a formula to describe the total number of tiles when you know the length of the path. You may use the formulas you found in problem **2.**

b. Explain your formula with a drawing.

4. a. Find another direct formula that also describes the total number of tiles when you know the length of the path.

b. Explain how this formula is different from the one you found in problem **3a.**

5. Use the formulas from problems **2–4** to answer these questions for a path with a length of 20:

a. How many white tiles are needed?

b. How many gray tiles?

c. How many total tiles?

Section B. Basic Patterns

The school store is having a sale on pencils, erasers, and rulers. Here is a list of the price of each item:

erasers	12 cents each
pencils	15 cents each
rulers	31 cents each

1. How much would a combination of three pencils, four rulers, and three erasers cost?

2. A teacher has her students work in small groups. She wants to buy two erasers, three pencils, and one ruler for each group.

a. How many items does each group get?

b. The teacher has six groups in her class. How much does it cost her to buy these items for each group?

c. Write a formula to find how much it costs to buy these items when you know the number of groups.

You can write a formula with a basic pattern. In our example, the number of items needed for each group is the basic pattern.

3. What is the basic pattern in terms of erasers, pencils, and rulers needed for one group?

4. Write a formula using this basic pattern to find how much it costs to buy items when you know the number of groups.

Four other teachers also need to buy items for the students in their classes. These teachers also want to give each of their groups two erasers, three pencils, and one ruler. The number of groups in each class is: 4, 7, 7, and 6.

5. How many items do all five teachers together need to buy?

6. How much does it cost to buy erasers, pencils, and rulers for all the groups in the five classes?

Section C. Area

Sal and his friends are going to carpet the floor of the tree house they are building. The floor of the tree house is a 6-by-6-meter square.

1. How many square meters is the area of the floor?

A local store has some cheap carpet remnants. Its carpet remnants are square. Sal and his friends have a list of carpet remnants from the store. Only the area is known for some of the remnants. Only the length of the side is known for other remnants. The size of the each remnant is listed below:

Piece I	area of 5 m^2
Piece II	area of 121 m^2
Piece III	side of 3 m
Piece IV	side of 3.5 m
Piece V	area of $\frac{1}{4}$ m^2

2. a. In your workbook, sketch the five pieces of carpet remnants.

b. Write down the length of a side and the total area for each piece. You may use the $\sqrt{}$ key on your calculator.

3. Which of these carpet remnants would you choose to cover the floor of the tree house?

Sal and his friends cannot find a combination from the available pieces to exactly cover the floor of the tree house. They telephone the carpet store and ask for more pieces. The store owner says he has a square remnant with an area of 11 square meters and one with an area of 23.75 square meters.

4. How long is the length of a side of the piece with an area of 11 square meters?

5. Now find two combinations of carpet remnants that exactly cover the floor of the tree house.

6. In your workbook, sketch how you would cover the floor using the available remnants. You may cut the pieces to make them fit the 6-by-6-meter area of the tree house floor.

Section D. Formulas

In Section D you learned that the faster you drive in a car, the longer the distance you need to stop your car. This is the stopping distance. Suppose the following formula is another possibility for finding the stopping distance:

$$\text{Speed (km/h)} \xrightarrow{\div 10} \_\_\_ \xrightarrow{\wedge 2} \_\_\_ \xrightarrow{\times 0.75} \text{Stopping Distance (m)}$$

1. What is the stopping distance when you are driving at the following speeds:

a. 100 kilometers per hour?

b. 50 kilometers per hour?

c. 120 kilometers per hour?

d. 80 kilometers per hour?

2. Rewrite the above formula as an equation. You may use letters to represent the speed and stopping distance.

3. If the stopping distance is 40 meters, how fast are you driving?

4. a. List in a table all the values you found in problems **1** and **3.**

b. Extend the table so it also lists a speed of 75 kilometers per hour and a stopping distance of 60 meters.

5. a. Make a graph to describe the relationship between speed and stopping distance. Put the speed on the horizontal axis and the stopping distance on the vertical axis.

b. Plot the points from the table in problem **4.**

c. Describe the shape of the graph.

6. Rewrite the formula so that you can find the speed when you know the stopping distance.

CREDITS

Cover

Design by Ralph Paquet/Encyclopædia Britannica, Inc.

Collage by Koorosh Jamalpur/KJ Graphics.

Title Page

Illustration by Brent Cardillo/Encyclopædia Britannica, Inc.

Illustrations

1–10, 12–13 Phil Geib/Encyclopædia Britannica, Inc.; **18–22** Paul Tucker/Encyclopædia Britannica, Inc.; **23** Brent Cardillo/Encyclopædia Britannica, Inc.; **24, 29–31, 33** Phil Geib/Encyclopædia Britannica, Inc.; **35–36, 38–39** Paul Tucker/Encyclopædia Britannica, Inc.; **40** Jerome Gordon; **42** Brent Cardillo/Encyclopædia Britannica, Inc.; **43–44** Phil Geib/Encyclopædia Britannica, Inc.; **47–48, 49 (top)** Jerome Gordon; **49 (bottom)** Phil Geib/Encyclopædia Britannica, Inc.; **52–54** Brent Cardillo/Encyclopædia Britannica, Inc.

Photographs

17 © Brent Cardillo/Encyclopædia Britannica, Inc.; **27** © D. Richter, Michigan Technological University.